Mechatronics: Engineering Fundamentals

Mechatronics: Engineering Fundamentals

Allie Weaver

MURPHY & MOORE
www.murphy-moorepublishing.com

Murphy & Moore Publishing,
1 Rockefeller Plaza,
New York City, NY 10020, USA

ISBN: 978-1-63987-358-6

Cataloging-in-Publication Data

Mechatronics : engineering fundamentals / Allie Weaver.
p. cm.
Includes bibliographical references and index.
ISBN 978-1-63987-358-6
1. Mechatronics. 2. Microelectronics. 3. Microelectromechanical systems.
4. Mechanical engineering. I. Weaver, Allie.

TJ163.12 .M43 2022

621--dc23

For information on all Murphy & Moore Publications
visit our website at www.murphy-moorepublishing.com

MURPHY & MOORE

Table of Contents

Preface

This book is a culmination of my many years of practice in this field. I attribute the success of this book to my support group. I would like to thank my parents who have showered me with unconditional love and support and my peers and professors for their constant guidance.

Mechatronics also known as mechatronic engineering is a field of engineering that deals with various types of mechanical and electrical systems. It is a multidisciplinary field and includes various other branches of engineering such as computer engineering, robotics, telecommunications, electronics, control, systems and product engineering. The goal of this discipline is to a make a design solution that integrates each of its subfields. It focuses on creating a simpler, reliable and economical system that integrates principles of both mechanics and electronics. Mechantronics finds its applications in a wide variety of areas such as servo-mechanics, industrial goods, computer aided and integrated manufacturing systems, medical imaging systems, structural dynamic systems and transportation and vehicular systems. This book is a valuable compilation of topics, ranging from the basic to the most complex theories and principles of this field. The topics included herein are of utmost significance and bound to provide incredible insights to the readers. This book will serve as a reference to a broad spectrum of readers.

The details of chapters are provided below for a progressive learning:

Chapter - Introduction

Mechatronics is the branch of engineering that is concerned with the engineering of both electrical and mechanical systems. A mechatronic system uses signals from the environment for generating output signals and transforms them into different forces and motions. This chapter delves into the subject of mechatronics for a thorough understanding of it.

Chapter - Digital Electronics

Digital electronics are referred to the types of electronics that operate on digital signals. It includes the concepts of number system, Boolean algebra, logic gates like AND, NOT, NAND, NOR, XOR and XNOR, combinatorial logic, sequential logic in synchronous and asynchronous circuit, etc. This chapter has been carefully written to provide an easy understanding of digital electronics.

Chapter - Analog Electronics

Analog electronics are the electronic systems that use continuous variable signals. Analogue filter, amplifier, bipolar junction transistor, field-effect transistor, etc. fall under its domain. The topics elaborated in this chapter will help in gaining a better perspective of the related aspects of analog electronics.

Chapter - Transducers

Transducers are the devices that convert a signal in one form of energy to another. A few of its types are active transducer, passive transducer, displacement transducer, pressure transducer, flow transducer, etc. All the types and concepts of transducers have been carefully analyzed in this chapter.

Chapter - Robotics

The interdisciplinary branch of engineering and science that is concerned with the design, operation, construction, and use of robots is referred to as robotics. It has a wide range of applications in the fields of medicine, agriculture, engineering, navigation and households. This chapter closely examines the field of robotics to provide an extensive understanding of the subject.

Chapter - Control System

A type of system that directs, commands, manages and regulates the behavior of different systems through control loops is termed as control system. It includes open and closed loop systems, feedback systems, PID control, etc. This chapter has been carefully written to provide an easy understanding of control system.

Allie Weaver

1
Introduction

Mechatronics is the branch of engineering that is concerned with the engineering of both electrical and mechanical systems. A mechatronic system uses signals from the environment for generating output signals and transforms them into different forces and motions. This chapter delves into the subject of mechatronics for a thorough understanding of it.

Mechatronics

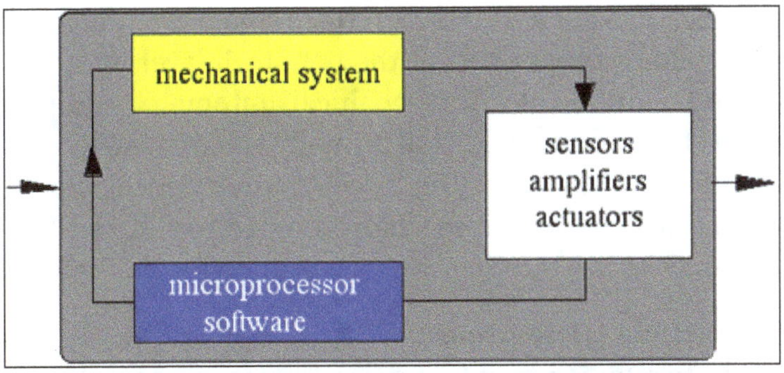

Mechatronics is an interdisciplinary area of engineering that combines mechanical and electrical engineering and computer science. A typical mechatronic system picks up signals from the environment, processes them to generate output signals, transforming them for example into forces, motions and actions.

It is the extension and the completion of mechanical systems with sensors and microcomputers which is the most important aspect. The fact that such a system picks up changes in its environment by sensors, and reacts to their signals using the appropriate information processing, makes it different from conventional machines.

Examples of mechatronic systems are robots, digitally controlled combustion engines, machine tools with self-adaptive tools, contact-free magnetic bearings, automated guided vehicles, etc. Typical for such a product is the high amount of system knowledge and software that is necessary for its design.

As a design philosophy, mechatronics values systems thinking and an interdisciplinary approach to problem solving. Fairly recently in information technology (IT), the DevOps movement broke down the wall between software development and operations, encouraging IT pros to work in cross-departmental teams. Mechatronics has caused the same sea change in engineering. In many industries today, engineers have to work in cross-discipline, collaborative teams in order to ensure that the complex, highly integrated systems they are designing will run smoothly.

To understand the concept of mechatronics, one can simply think about today's automobiles. The average car today has between 25 and 50 central processing units (CPUs) that control mechanical functions. A driver support system (DSS) such as anti-lock brakes is designed with mechatronics; the electronic control system takes over the braking function when sensors recognize that one or more wheels are locking up. A pneumatic tire pressure monitoring system is also designed with mechatronics; each tire has a sensor inside that sends data to an onboard electronic control system. If the pressure on one tire is low, the embedded software in the control system sends an alert to the vehicle's dashboard and a tire gauge icon lights up. Even an automobile's air bag is designed with mechatronics; a micro-electrical machine (MEM) in the front of the automobile will deploy an airbag when sensor data indicates rapid deceleration.

Thinking holistically has allowed engineers to make great strides in advancing automation and robotics. It has also allowed them to create smarter products, smaller products and tiny, efficient product components at the nanotech level. In order to successfully operate, diagnose and maintain electromechanical, automated equipment, it's become necessary for support technicians to re-skill formally and informally on a frequent basis. Siemens AG, one of the world's largest high-tech manufacturing corporations, offers a professional certification program for mechatronics, and an increasingly large number of colleges and universities are offering courses of study for this emerging technology.

According to the United States Department of Labor, mechatronics is an emerging growth area for employment. In addition to automotives, industries that rely upon mechatronics include aerospace, appliance design and repair, banking, energy and power production and distribution, farming, food processing, manufacturing, mining and healthcare.

Mechatronic System

A mechatronic system can be characterized synthetically by versatility and flexibility as two main aspects whose integration gives mechatronic behaviour of modern systems.

Versatility is mainly related to operation capability and performance that are needed for mechanical activity of a mechatronic system. Flexibility is mainly related to regulation capability and performance that are needed for controlled actions of a mechatronic system.

Thus, mechanical issues in mechatronic systems are related to versatility mainly in terms of Kinematics and Dynamics of a load movement, Mechanics of interactions (like contact and grasp), Dynamics of Multibody Systems. However, the mechanical attention is always linked to aspects of the mechatronic design and indeed, simulation of mechanical actions is usually performed by taking into account also models for the controlled actuation.

Thus, the role of Mechanical Engineering in Mechatronics can be understood according to two main aspects, namely the mechanical design and operation, and the mechanical interaction with environment in performing system tasks.

Those mechanical aspects both for design and operation aims are studied by looking at traditional disciplines for machinery but even at specific novel disciplines. Thus, Kinematics and Dynamics of load movement are studied to analyze and investigate on the motion of mechatronic systems and load body during the operation performing or not a task.

Particular attention is usually directed to mechanical aspects of the motion in the system and load as due to the flexibility and control of the operation of the system. In addition the task is studied by looking both at kinematic and dynamic feature of the motion and related actions against the environment and within the mechatronic system yet. The particular attention to motion issues is also motivated by the attention to safety and security issues both for the system and human operators that can be in the operation area of the system.

Mechanics of interaction is interesting in evaluating situations with mechanical contacts and force transmissions between the system parts or its extremity and environment or task object. It is fundamental to size the system actions according to the task requirements and specific analysis of the corresponding mechanics is necessary to achieve desired goals and proper working of the overall system.

Dynamics of Multibody systems is a mechanical discipline that is oriented to consider complex systems during complex motions like spatial movement at high acceleration and velocity, with the aim to take into account any inertia effect. Indeed, this discipline has been developed by looking at integrated systems and very often the dynamics of Multibody systems include suitable modelling of components of other nature than pure mechanical one, like for example servo-controlled actuators whose motion and dynamics can strongly affect the system behaviour and operation characteristics, as it is mechatronic systems.

Mechanical issues in mechatronic systems can be considered for specific goals and operation tasks, like for example locomotion mobility, object grasping and manipulation, force transmission, and so on.

Versatility of a mechatronic system can be seen and understood under different aspects, and the above-mentioned considerations can be related to main aspects which make the role of Mechanical Engineering fundamental for the success of the mechatronic systems, but within a frame of multidisciplinary integration yet.

Mechatronic Design System

Mechatronic design is the integrated design of a mechanical system and its embedded control system. In order to make proper choices early in the design stage, tools are required that support modeling and simulation of physical systems–together with the controllers–with parameters that are directly related to the real-world system.

The integration within a mechatronical system is performed through the combination of hardware (components) and software (information processing). Hardware integration results from designing the mechatronic system as an overall system and bringing together the sensors, actuators, and microcomputers into the mechanical system. Software integration is primarily based on advanced control functions.

The first step in the focused development of mechatronical systems is to analyze the customer and the technical environment in which the system is integrated. Complex technical systems designed to solve problems tend to be a combination of mechanical, electric, fluid, power, and thermodynamic parts with hardware in digital and analog form coordinated by complex software. Typical mechatronic systems gather data and information from their technical environment using sensors. The next step is to use elaborate ways of modeling and description methods to cover all subtasks of this system in an integrated manner. This includes an effective description of necessary interfaces between subsystems at an early stage. The data are processed and interpreted, leading to action carried out by actuators. Mechatronical systems result in shorter developmental cycles, lower costs, and higher quality. They also provide additional influence through the acquisition of information from the process.

Mechatronic design supports the concepts of concurrent engineering. In the designing of a mechatronic product, it is necessary that the knowledge and required information be coordinated among different expert groups. Concurrent engineering is a design approach in which the design of a product and manufacture of a product are merged in a special way. Traditional barriers between design and manufacturing are removed. It has been influenced partly by the recognition that many of the high costs in manufacturing are decided at the product design stage itself. Even during the design stage, it is involved with customer perception, market analysis, optimized performance, life cycle performance, quality, reliability, and sales. Product design and planning take place concurrently. The total philosophy of concurrent engineering in the organization is well suited for team-oriented project management, with emphasis on collective decision making. Successful implementation of concurrent engineering is possible by coordinating adequate exchange of information and dealing with organization barriers to cross-functional cooperation. Due to the influence of concurrent engineering, traditional barriers between design and manufacturing have decreased; however, the lack of a common interface language has made the information exchange in concurrent engineering difficult.

A mechatronic product can achieve impressive results if it is effectively integrated with the concurrent engineering management strategy. The benefits that accrue are greater productivity, higher quality, and production reliability by the incorporation of intelligent, self-correcting sensory feedback systems. The integration of sensors and control systems in a complex system reduces capital expenses, maintains a high degree of flexibility, and results in a higher percentage of machine utilization. In order to implement a mechatronic concurrent engineering system that can achieve these objectives, the organization must start with a long-range plan that can take into accounts tomorrow's changing needs in process, data functions, control, and integration tools.

The mechatronic design methodology is concerned not only with producing high-quality products but with maintaining them as well, an area referred to as life cycle design.

Several important life cycle factors are:

- Delivery: Time, cost, and medium.
- Reliability: Failure rate, materials, and tolerances.
- Maintainability: Modular design.
- Serviceability: On board diagnostics, prognostics, and modular design.
- Upgradeability: Future compatibility with current designs.
- Disposability: Recycling and disposal of hazardous materials.

In the mechatronic design approach, life cycle factors are included during the product design stages, resulting in products that are designed from conception to retirement. The mechatronic design process is presented in figure.

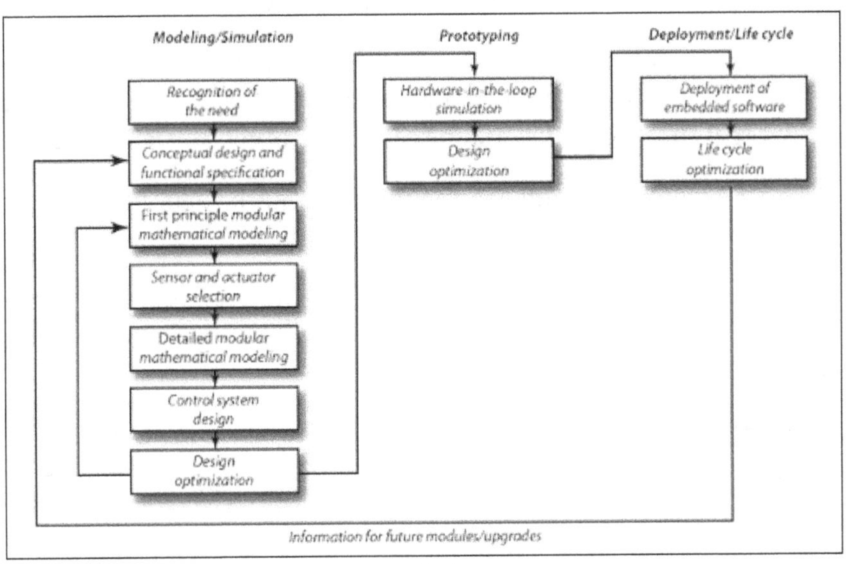

Mechatronic design process.

The mechatronic design process consists of three phases: modeling and simulation, prototyping and deployment. All modeling, whether based on first principles (basic equations) or the more detailed physics, should be modular in structure. A first principle model is a simple model that captures some of the fundamental behavior of a subsystem. A detailed model is an extension of the first principle model providing more function and accuracy than the first-level model. Complex models may be created by connecting the modules, or blocks, together.

Each block represents a subsystem that corresponds to some physically of functionally realizable operations that can be encapsulated into a block with input/output limited to input signals, parameters, and output signals. Of course this limitation may not always be possible or desirable; however, its use will produce modular subsystem blocks that can be easily maintained, exercised independently, substituted for one another (first principle blocks substituted for detailed blocks and vice versa), and reused in other applications. Following deployment, the life cycle functions begin, and information from these functions should be made available for upgrades to the product.

Because of their modularity, mechatronical systems are well suited for applications that require reconfigurations. Such products can be reconfigured either during the design stage by substituting various subsystem modules or during the life span of the products. Since many of the steps in the mechatronic design processes rely on computer-based tasks such as information.

Fusion, management, and design testing, and efficient computer-aided prototyping environment is essential. Some of the important features of such an environment are:

- Modeling: Block diagram or visual interface for creating intuitively understandable behavioral models of physical or abstract phenomena. The ability to encapsulate complexity and maintain several levels of sub model complexity is useful.
- Simulation: Numerical methods for solving models containing differential, discrete, hybrid, partial, and implicit nonlinear se well as linear equations. Must have a lock for real-time operation and be capable of executing faster than real time.
- Project Management: Database for maintaining project information and sub models for eventual reuse.
- Design: Numerical methods for constrained optimization of performance functions based on model parameters and signals. Monte Carlo type capability is also desirable.
- Analysis: Numerical methods for frequency domain, time domain, and complex domain design.
- Real-Time Interface: A plug-in card is used to replace part of the model with actual hardware by interfacing to it with actuators and sensors. This is called a

hardware-in-the-loop simulation or rapid prototyping and must be executed in real time.

- Code Generator: To produce efficient high-level source code from the block diagram or visual modeling interface. The control code will be compiled and used on the embedded processor. The language is usually C.
- Embedded Processor Interface: The embedded processor resides in the final products, and this feature provides communication between it and the computer-aided prototyping environment. This is called a full system prototype.

Because no single model can ever flawlessly reproduce reality, there will always be error between the behavior of a product model and the actual products. These errors, referred to as unmodeled errors, are the reason why so many model-based designs fail when deployed top the product. The mechatronic design approach also uses a model-based approach, relying heavily on modeling and simulation; however, unmodeled errors are accounted for in the prototyping step. Their effects are absorbed into the design, which significantly raiser the probability of successful product deployment.

In the prototyping step many of the non-computer subsystems of the model are replaced with actual hardware. Sensors and actuators provide interface signals necessary to connect the hardware subsystems back to the model. The resulting model is part mathematical and par real. Because the real part of the model inherently evolves in real time and the mathematical part evolves in simulated time, it is essential that the two parts be synchronized. This process of fusing and synchronizing model, sensor, and actuator information is called real-time interfacing or hardware-in-the-loop simulation and is and essential ingredient in the modeling an simulating environment.

Advanced approaches in Mechatronics

In addition to influencing the way the products are designed, recent developments in mechatronics are also creating opportunities in intelligent manufacturing. In order to produce high-quality products at low life cycle costs, many manufacturing plants have been reducing their workforce. As a result, the reduced workforce has to cope with additional tasks and responsibilities.

And there is a broader area of control for plant management. This trend has significantly influenced the mechatronic field, especially in the area of sensor-based manufacturing. It has created opportunities for the use of intelligent autonomous inspection systems as well as intelligent decision-making systems that perform tasks automatically, without human intervention, by adapting to the changes in the environment.

Model-based monitoring using intelligent sensors is an important technology holding much promise for improving manufacturing productivity and quality. In model-based monitoring the model is used to represent the ideal behavior of the structure. It is driven with the same inputs used for the process. The difference between the model output and the actual process output signals provides a concise mechanism for incorporating diagnostics, an attractive alternative to empirical rule-based decisions systems.

Monitoring systems can also be used to maintain machine and process performance during the manufacturing process. Figure presents a generic diagram of an intelligent model based manufacturing system.

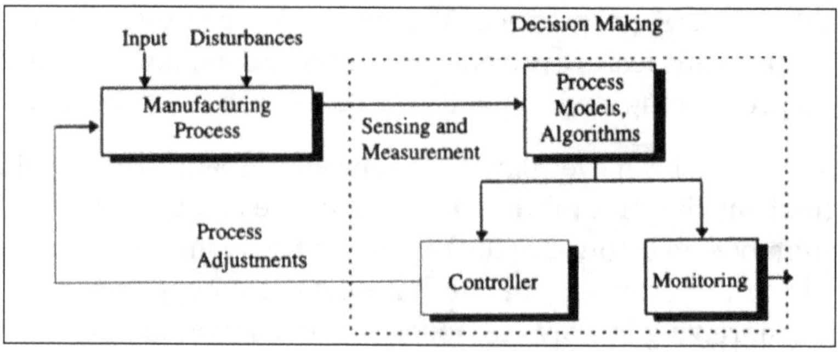

Model-based monitoring system.

In Figure the monitoring output indicates the status of the monitoring process. The controller applies commands to the process such that various sensed values, related to machine and/or process performance, are maintained, or regulated, at desired values.

Mechatronic technology used in manufacturing will impact new equipment as well as some retrofit applications. Most of the obstacles that are encountered in retrofit applications are due to hardware incompatibility and closed software structures. Computerized Machining Centers and Programmable Logic Controllers often utilize proprietary hardware, making integration between different systems difficult. The closed structure of the real-time process control, often found in older machinery, also presents implementation problems.

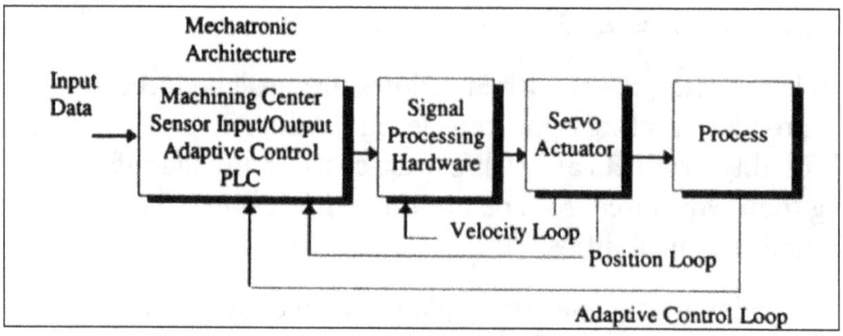

Mechatronic system with open-architecture platform.

The development of modular, open-architecture machine controllers, and figure can provide improvements to the existing system and enhance flexibility. The growth of open architecture systems for machine tool control will result in greater utilization of real-time process monitoring and control modules. This trend is also responsible for accelerating the use of systems will require a system perspective on manufacturing problems. A system approach which considers various solution alternatives for obtaining desired production performance will likely employ mechatronic technology.

Examples of mechatronical systems for industrial use are found in many areas. Mechatronic monitoring systems have been applied to products such as aircraft, machine tools, and automobiles. These systems are designed to measure plant parameters such as compliances and inertias, plant states such as current and velocity, and production states such as force and wear.

Some other noteworthy mechatronic applications are:

- Autonomous production cells with image-based object recognition.
- Monitoring and control of the welding process (currently under research at the European Center of Mechatronics).
- Mechatronical systems with off-line programming that automatically program a robot after receiving CAD data.
- Integrated supervisory systems with multi-process control capability and shared databases from the CD drawings.
- Flexible manufacturing systems with off- and on-line programming.
- Biorobotics, which utilize the bio functions for applications in environmental control.
- Endoscopic and orthopedic surgery.
- Magnetically levitated vehicles.
- Robotics in nuclear inspection and space applications.

Numerical computation, simulation, computer-aided design and experimental validation are important technologies that must be considered when evaluating the feasibility of complex mechatronical systems. Other technologies include artificial intelligence, expert systems, fuzzy logic, and neural networks. The usefulness of these technologies is expected to be at the higher levels of the control hierarchy in machining processes.

Manufacturing

Importance of Mechatronics in Automation

Today's customers are demanding more variety and higher levels of flexibility in the products. Due to these demands and competition in the market, manufacturers are thriving to launch new/modified products to survive. It is reducing the product life as well as lead-time to manufacture a product. It is therefore essential to automate the manufacturing and assembly operations of a product. There are various activities involved in the product manufacturing process. These activities can be classified into two groups viz. design and manufacturing activities.

Mechatronics concurrently employs the disciplines of mechanical, electrical, control and computer engineering at the stage of design itself. Mechanical discipline is employed in terms of various machines and mechanisms, whereas electrical engineering as various electric prime movers viz. AC/DC, servo motors and other systems is used. Control engineering helps in the development of various electronics based control systems to enhance or replace the mechanics of the mechanical systems. Computers are widely used to write various software to control the control systems product design and development activities; materials and manufacturing resource planning, record keeping, market survey, and other sales related activities.

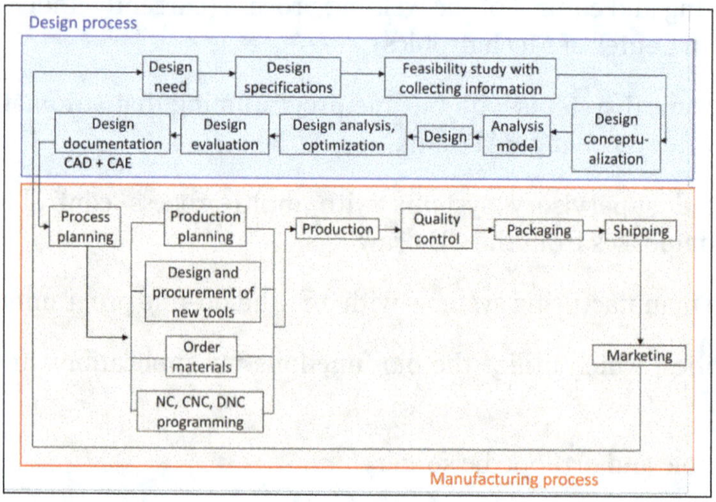

Operations involved in design and manufacturing of a product.

Using computer aided design (CAD)/computer aided analysis (CAE) tools, three-dimensional models of products can easily be developed. These models can then be analyzed and can be simulated to study their performances using numerical tools. These numerical tools are being continuously updated or enriched with the real-life performances of the similar kind of products. These exercises provide an approximate idea about performance of the product/system to the design team at the early stage of the product development. Based on the simulation studies, the designs can be modified to achieve better performances. During the conventional design manufacturing process, the design assessment is generally carried out after the production of first lot of the products. This consumes a lot of time, which leads to longer (in months/years) product development lead-time. Use of CAD–CAE tools saves significant time in comparison with that required in the conventional sequential design process.

CAD-CAE generated final designs are then sent to the production and process planning section. Mechatronics based systems such as computer aided manufacturing (CAM): automatic process planning, automatic part programming, manufacturing resource planning, etc. uses the design data provided by the design team. Based these inputs, various activities will then be planned to achieve the manufacturing targets in terms of quality and quantity with in a stipulated time frame.

Mechatronics based automated systems such as automatic inspection and quality assurance, automatic packaging, record making, and automatic dispatch help to expedite the entire manufacturing operation. These systems certainly ensure a supply better quality, well packed and reliable products in the market. Automation in the machine tools has reduced the human intervention in the machining operation and improved the process efficiency and product quality. Therefore it is important to study the principles of mechatronics and to learn how to apply them in the automation of a manufacturing system.

Mechatronics System

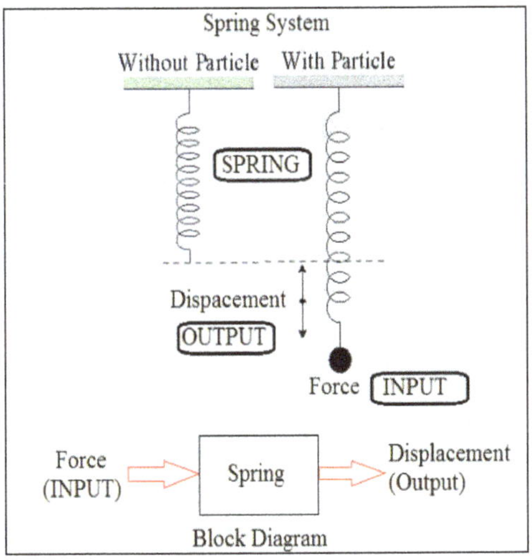

A spring-force system.

A system can be thought of as a box or a bounded whole which has input and output elements, and a set of relationships between these elements. Figure shows a typical spring system. It has 'force' as an input which produces an 'extension'. The input and output of this system follows the Hooke's law $F = -kx$, where F is force in N, x is distance in m and k is stiffness of the spring.

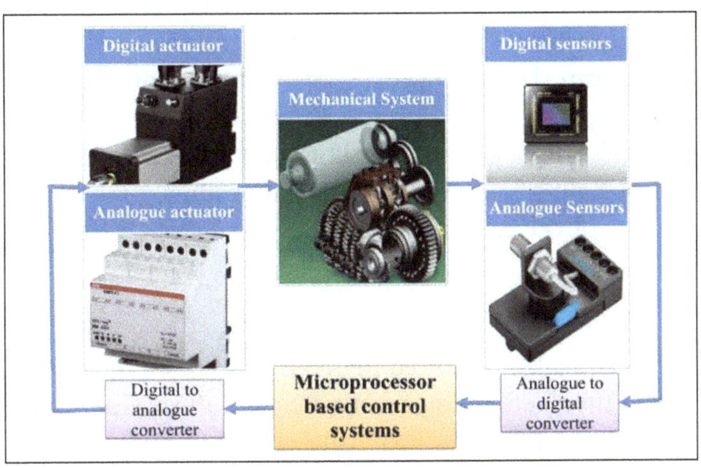

Constituents of a mechatronics system.

A Mechatronics system integrates various technologies involving sensors, measurement systems, drives, actuation systems, microprocessor systems and software engineering. Figure above shows the basic elements of a mechatronics system. To replace the mechanics of this mechanical system with an equivalent mechatronics based system, we need to have the basic controlling element, a microprocessor. Microprocessor processes or utilizes the information gathered from the sensor system and generates the signals of appropriate level and suitable kind (current or voltage) which will be used to actuate the required actuator viz. a hydraulic piston-cylinder device for extension of piston rod in this case. The microprocessor is programmed on the basis of the principle of Hooks' Law. The schematic of microprocessor based equivalent spring mass system is shown in figure below:

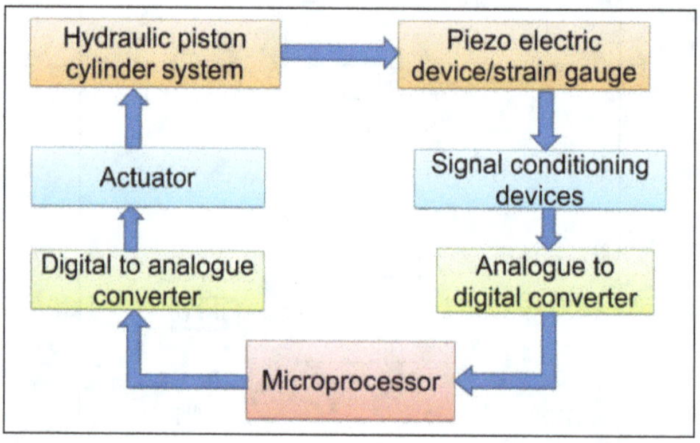

Microprocessor based equivalent spring mass system.

The input to the system is a force which can be sensed by suitable electro-mechanical sensors viz. piezo-electric device or strain gauges. These sensors generate either digital signals (0 or 1) or analogue signals (milli-volts or milli-amperes). These signals are then converted into right form and are attenuated to a right level which can properly be used by the microprocessor to take generate the actuation signals. Various electronics based auxiliary devices viz. Analogue-to-Digital Converter (ADC), Digital-to-Analogue Converter (DAC), Op-amps, Modulators, Linearization circuits, etc. are used to condition the signals which are either received by the microprocessor from the sensors or are sent to the actuators from the microprocessor. This mechatronics based spring-mass system has the input signals in the digital form which are received from the ADC and Piezo-electric sensor. The digital actuation signals generated by the microprocessors are converted into appropriate analogues signals. These analogue signals operate the hydraulic pump and control valves to achieve the desired displacement of the piston-rod.

Servo Mechanical Systems

Mechatronic systems typically exhibited high a degree of complexity due to the strong cross coupling of the involved different engineering disciplines such as mechanical, electronic and computer. This complexity originates from the large number of couplings on various levels of the contributing elements and components, coming from different disciplines. The difficulty for the design engineer in his daily work is that these couplings have to be considered in an early phase of the design process. With shortening product lift cycle, design managers are consistently trying to identify means for producing a better product in a shorter period of time.

Therefore, the realm of Mechatronics is high speed, high precision, and high efficiency, highly robust. The difficulty in the Mechatronic approach is that it requires a system perspective: system interactions are important, system modeling is required, and feedback control systems can go unstable. Mechatronic design concepts include direct-drive mechanisms, simple mechanics, system complexity, accuracy and speed from controls, efficiency and reliability from electronics, and functionality from microcomputers. Starting at design and continuing through manufacture, Mechatronic designs optimize the available mix of technologies to produce quality precision products and systems in a timely manner with features the customer wants. The real benefits to industry of a Mechatronic approach to design are shorter development cycles, lower costs, and increased quality, reliability, and performance.

Additionally, in order to evaluate concepts generated during the design process, without building and testing each one, the mechatronics engineer must be skilled in the modeling, analysis, and control of dynamic systems and understand the key issues in hardware implementation. Thus, as the figure shows, the essential characteristic of a mechatronics engineer and the key to success in mechatronics system is a balance between two sets of skills.

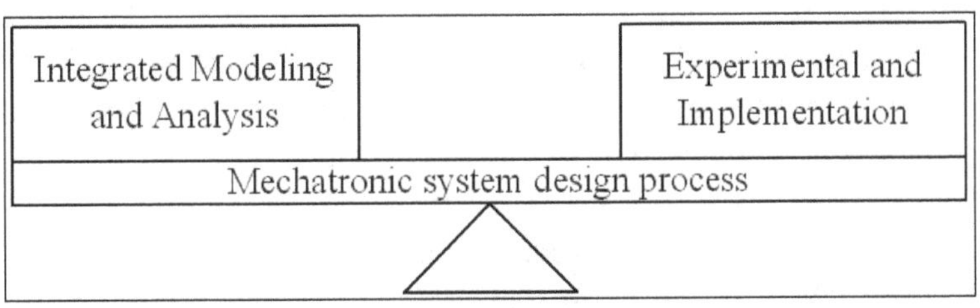

Balance of mechatronic design process.

Integrated Modeling and Analysis of Dynamic Mechatronic Systems

During the design of mechatronic systems, it is important that changes in the mechanical structure and the controller be evaluated simultaneously. Although a proper controller enables building a cheaper mechatronic system, a badly designed mechanical system will never be able to give a good performance by adding a sophisticated controller. Therefore, it is important that during an early stage of the design a proper choice can be made with respect to the mechanical properties needed to achieve a good performance of the controlled system. On the other hand, knowledge about the abilities of the controller to compensate for mechanic imperfections may enable that a cheaper mechanical structure be built. This requires that in an early stage of the design a simple integrated model is available, that reveals the performance limiting factors of the mechatronic system.

Consequently, in order to help mechanical structure and controller of mechatronic system modeling simultaneously, the mechatronic system design methods must be integration. Accordingly, some of numerical based integrated design strategies for mechatronic system were proposed to some fields such as: aerospace, robotics and manufacturing systems in the early years. However, the dynamic models derived with the above integrated methods typically have a high order. A critical issue in the mechanical structure and control modeling with the integrated design approach is difficulty from each domain.

Therefore, for complex multibody systems of mechatronics, graphical modeling software is helpful to formulate automatically the equations of motion from a high-level description. Among the computer modeling methods, symbolic methods allow to build the equations of motion in symbolic format, whereas numerical methods produce the equations of motion as complex numerical procedures. The symbolic format has the advantages of portability and efficiency, and it provides interesting insights in the analytical structure of the equations. However, numerical methods are able to deal with a more general class of problems, and they are especially suitable to model the dynamics of a flexible mechanism with complex topology in a systematic way.

Experimental Validation and Hardware Implementation of Designs

In an industrial process, design of controllers involve formulation of reasonably accurate models of the plant to be controlled, designing control laws based on the derived models and simulating the designed control laws using available simulation tools such as MATLAB/Simulink. Whereas implementation is accomplished by converting the designed control laws to the native code of target systems, most commonly embedded microprocessor based architecture or personal computer with analog and digital interfaces. Controllers can be designed in the continuous, discrete or hybrid time domain whereas implementation is accomplished mostly in discrete time domain as most of the present day controllers are being implemented in digital machines. Presence of the vast difference in design and implementation of control applications is inherent due to different concepts in the field of control engineering and computer science. Thus, transformation of controller designs to implementation induces possibilities of errors and unreliable behaviors. In some cases, these errors cannot be identified by rigorous tests of the implementation thus these errors results in failure of the system causing serious and even catastrophic disaster.

Furthermore, the typical controller design task requires selection of controller strategies, structures and parameter values. Before implemented engineers should be tested using actual plant data or in prototype implementation with physically measured inputs and generated outputs. This phase is necessary for experimental validation of model simplifications and other assumptions made when designing the controller. On the other hand, real-time simulation provides the best conditions for performance tuning. However, sometimes the reverse situation occurs when plant model is substituted for the actual plant while the controller might be fully implemented. This approach is called Rapid Controller Prototyping (RCP) simulation. For this technique, engineers have actuators, sensors and other physical components interfacing with real-time simulation. Furthermore, RCP techniques allow implementing and validating control strategies during the development process that users can work within the same environment from the requirement analysis to the controller design and implementation phase.

According to those two sets of skills, in the mechatronic design process, it can be broadly categorized into three stages in a computer-supported design environment, namely, the design problem of understanding behavior of mechatronic system through an analysis of need, initial solution generation through conceptual design, and solution refinement and finalization through multi-discipline detailed design. In computer support for engineering design, there is little support for the first two stages in the design process, primarily due to the complexity and diverse needs of these design activities during the three stages. The final stage in the design process is currently the main area that has reasonable computer support, and can be used to assist engineer designers to improve their designs or products. This stage of computer support can be further decomposed into component modeling, component matching and sizing, and behavior simulation and comparison for informative decision-making. This decomposition facilitates further investigation of the constituents of each design support activity.

Notably, one typical problem with many current computer-modeling methods is that they are extremely domain dependent. In the mechatronic system design processes, which include structural design, controller design and implementation in three domains, also consider interactions among multiple domains, such as integrated design, rapid prototyping and animation technology. Therefore, mechatronic design engineers must to be trained to use in different application domains such that they would be competent in using all these domain-dependent technologies. This task itself is very challenging. Consequently, to solve dependent problems for mechatronic systems, mechatronic engineers always use a dynamic equation that includes all parameters in the structural and control domains. Unfortunately, one of the most significant problems when using equation-based mechatronic modeling is the amount of modeling data that must be analyzed during the process. Because of this enormous amount of data and the numerical algorithm that must also be utilized, this method of modeling and simulation is typically very slow and prone to errors. Additionally, this method requires excellent knowledge of numerical solution methods and programming principles.

Based on those reasons, in order to easy integrated design and simulation of mechatronic system for different concept domains, in this study the graphical environment called Computer Aided Rapid System Integration (CARSI) technology will be developed to achieve the structure design, controller design and implementation in the same design environment.

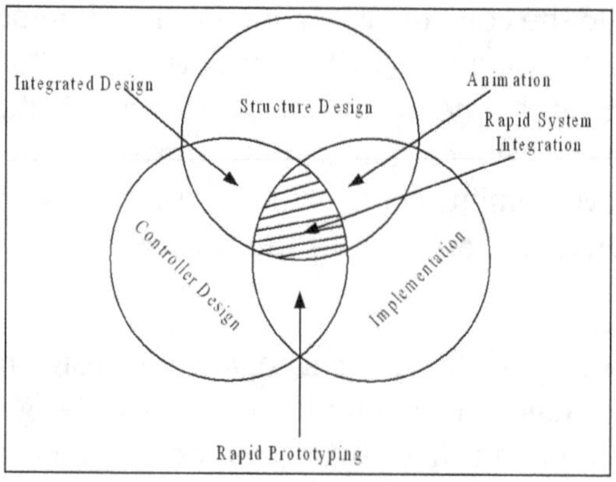

Skills for mechatronic system design.

Integrated Design Strategy

With a multilevel decomposition approach, a large complex optimization problem is broken into a hierarchy of smaller optimization sub problems. This hierarchy can be thought as levels of increasing details. At the upper level, the sub problem is formulated in terms of global quantities, which describe the overall behavior of the entire system. On the lower level, the sub problems are stated in terms of local quantities and local constraints, which have only a small impact on the entire system. Each sub problem uses local design variables to reduce the violation of constraints, which are unique to that sub problem. Each level is a multi-objective optimization problem characterized by a vector of objective functions, constraints and design variables. So considering the structure and control two-level problem for a mechatronic system, the multilevel decomposition procedure can be written as below.

At structure level,

$$\text{min. } Y_{Ni}(X_N), i=1,....,n_N$$

$$\text{s.t. } g_{Nk}(X_N) \le 0, k=1,....,n_{cN}$$

$$\sum_{i=1}^{NDV_N} \frac{\partial Y_{Rj}}{\partial X_{Ni}} \Delta X_{Ni} \le \varepsilon_{2j}, j=1,......,n_R$$

$$X^L_{N_i} \le X_{Ni} \le X^U_{Ni}, i=1,......,NDV_N$$

$$X^L_{Rj} \le X^*_{Rj} + \sum_{i=1}^{NDV_N} \frac{\partial X^*_{Rj}}{\partial X_{Ni}} \Delta X_{Ni} \le X^U_{Rj}, j=1,...,NDV_R$$

where Y_N and Y_R are the objective function vectors at the structure level and the control level, respectively; g_N and g_R are the corresponding constraint vectors; X_N and X_R are the corresponding design variable vectors, ε_{2j} is a tolerance on the change in the j_{th} objective of control level during optimization at the structure level; L and U are lower and upper bounds of design vectors, $\partial Y^*_{Rj} / \partial X_{Ni}$ and $\partial X^*_{Rj} / \partial X_{Ni}$ represent the optimal sensitivity parameters of the control level objective function and design variable vectors, respectively, with respect to the structure level design variables. n_N and n_R denote numbers of objective functions for each level; n_{cN} is the number of constraints for structure level; NDV_N and NDV_R are numbers of design variables for the structure and the control levels.

Similarly, the process of control level becomes,

$$\min\ Y_{Rj}(X_N^*, X_R), j = 1, \ldots, n_R$$
$$\text{s.t.}\quad g_{Rk}(X_N^*, X_R) \le 0, k = 1, \ldots, nc_R$$
$$X_{Ri}^L \le X_{Ri} \le X_{Ri}^U, i = 1, \ldots, NDV_R$$

Where X_N* is the optimum design variable vector from the structure level and must be fixed during optimization at the control level.

Following min. $Y_{Ni}(X_N), \ldots NDV_R$ and min $Y_{Rj}(X_N^*, X_R) \ldots, NDV_R$ the integrated design methodology can be broken into sequential, iterative, and simultaneous three strategies:

In the sequential strategy, the mechanical structure is usually designed first. It is then fitted with off-the-shelf electric motors and drive electronics. Finally, a controller is designed and tuned for the existing physical system until the goal is archived therefore, it is called Design Then Control (DTC) strategy. In this method, the structure is assumed to be fixed and cannot be changed by excluding considerations from a dynamics and control point of view. Consequently, this approach leads to a system with non-optimal dynamic performance.

For this method, the structural design is also first performed based on loading considerations. Sizes and masses of mission-related components are estimated and a structure that maintains the desired component relationships during operations is designed. Next, a controller is designed for the fixed structure to obtain the required dynamic performance. The control design must also provide satisfactory closed-loop stability and robustness properties. If the nominal system does not provide an adequate performance, the design process must return to the structural discipline for modification. After modification, the structure parameters are returned to the control discipline for redesign. This iterative process continues until a satisfactory compromise is found between the mission and control requirements. Now suppose that it is desired to simplify the min. $Y_{Ni}(X_N), \ldots NDV_R$ and min $Y_{Rj}(X_N^*, X_R) \ldots, NDV_R$ formulation as much as possible. One could presumably simplify the problem by assuming that all the objective functions and constraints are convex within both the structure and controller design subspaces. In other words, one could presumably assume that, when the structure design variables X_N are fixed, all the objective functions and constraints in the above problem will be convex and vice versa. However this assumption is not a sufficient guarantee for the system level optimization problem to be convex. Thus, in order to achieve the optimization problem into the system level, the simultaneous design strategy must be considered.

As min. $Y_{Ni}(X_N)$,...NDV_R and min $Y_{Rj}(X_N^*, X_R)$...,NDV_R, given a combined structure and controller optimization problem for mechatronic system, the system level is often nonconvex, even if the individual structure and control optimization sub-problems are convex (individual design problem for min. $Y_{Ni}(X_N)$,...NDV_R and min $Y_{Rj}(X_N^*, X_R)$...,NDV_R. The main reason is easy involved the static and variation optimization problem during iterative design process.

Therefore, as figure shows, comparing above three strategies, even system performance will be increased during sequential, iterative, and simultaneous strategies.

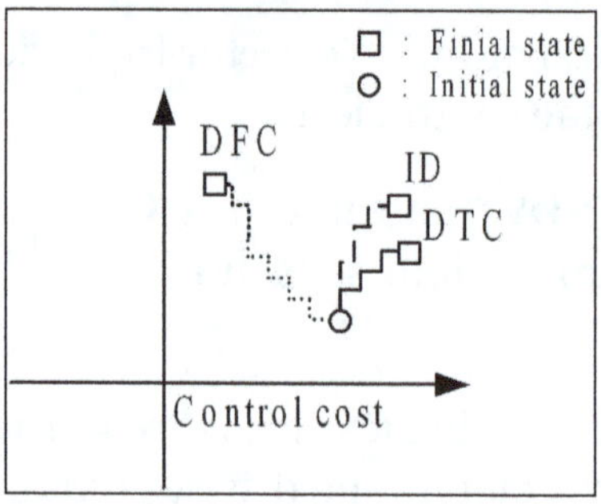

Control cost in iterative process.

Construction Automation

Sensing

The 3 D measurements are useful for autonomous mobile robot to recognize its environments as well as the reconstruction of 3 D information of large scale structures in construction sites. The combination of a LRF and a camera can reconstruct the inside of a building. Images from the camera are mapped on the corresponding 3D scan data from LRF by using texture mapping technique as shown in figure.

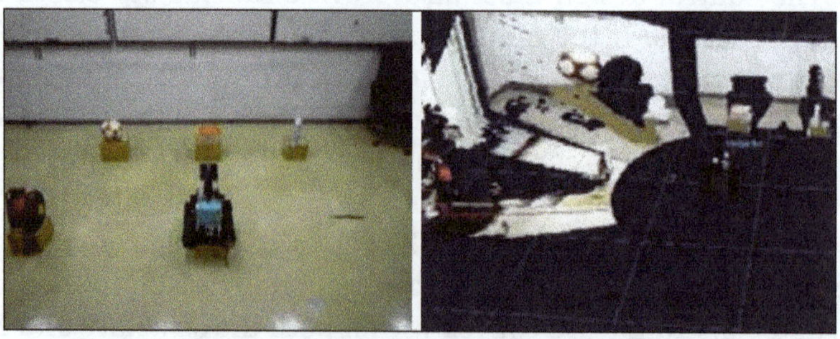

(a) Robot & environment. (b) Reconstructed 3D mapping.
Combination of LRF and camera data for 3D map.

The SLAM is useful technique to achieve map building and robot localization at the same time. The unique 6-DOF SLAM by using scan matching based on the 3-D distance information acquired by omnidirectional stereo vision. Acquired scan is evaluated by the Iterative Closest Point (ICP) algorithm, and the result is integrated by the extended Kalman filter.

Actuation

The dexterous hand mechanisms are capable of manipulating and handling various objects in construction sites. A unique finger mechanism is proposed with omnidirectional driving roller to realize the two active rotational axes on the surface of the grasped object. The cylindrical tracked unit can be applied as roller and the fingers can manipulate the grasped object in the arbitrary axes as in figure.

Omnidirectional finger.

Another example is omnidirectional enfolding gripper which is capable of grasping various objects with deformable bag as shown in figure which contains any materials switching coagulation and dissolution phases, e.g. media (gas, liquid) and particles, functional fluid, etc. Changing its mode from liquid-like state to solid-like can enlarge grasping force and the grip can hold even complicated shape objects stably.

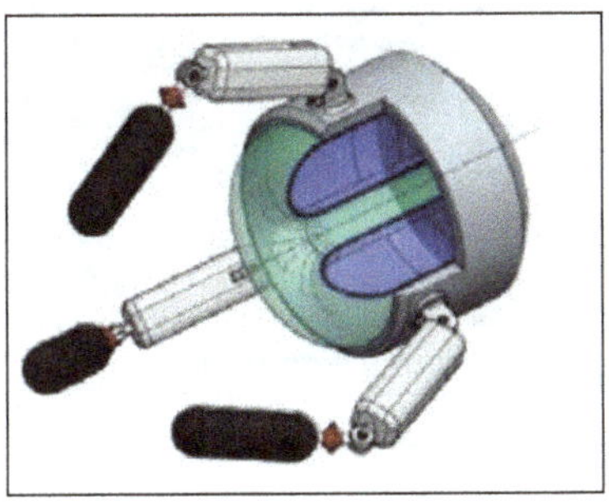

Enfolding gripper.

Energy saving in power supply is another crucial aspect in the on-site construction. A variable volume tank is capable of driving pneumatic actuator with low discharged pressure in a tank.

Mobile Robot

Mobile manipulation is one of the active research areas. A new mobile manipulator platform was designed and built in the NEDO Intelligent Robot Project to promote development of so-called intelligent robot modules (RTM) for generic robot motion programs and application tasks. The omnidirectional mobile capability is useful in complicated construction environments. The mobility can be more enhanced by introducing an active caster mechanism. The active-caster wheel features simple mechanism and high positioning accuracy, and will allow a robot to carry out more accurate object transportation.

Development of leg robot is still active and expected in the rough terrain tasks as well as in narrow and small spaces in the inside of buildings and large structures for inspection and monitoring tasks.

ASTERISK in monitoring.

The crawler mechanism is promising mobile platform to achieve high-speed stable locomotion on rough terrain. Many prototypes have been proposed relating to rescue operations. The recently developed platform, "Quince" is expected to work in the inside of Fukushima NPP for monitoring and light duty tasks.

Quince on rough.

Power Assist

The power assist system has its long history in its development starting in early 1960s at General Electric. The recent advancement of the computer technology has brought a chance to realization of practical power assist mechanism. There are more than 15 research activities in Japanese universities. The main objective is to support weakened senior people and the disabled in their daily life as well as to assist workers in heavy duties which are common in construction sites. The key issues are compact actuation mechanism design, human interface, portable energy source, and so on.

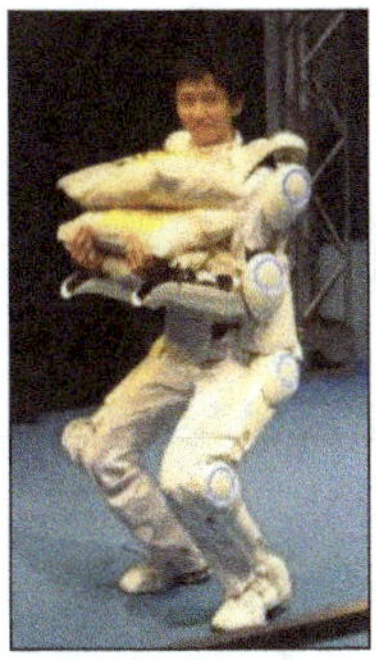 

HAL. Muscle suit.

Humanoid

The humanoid research and development are one of the most active robotics in Japan. HONDA ASIMO, TOYOTA Partner Robot, and AIST HRPs are well established humanoid platforms. The HRP2 is used as a common platform in universities and institutes where they carry out fundamental researches such as stable walking with energy consumption, practical use of vision for object recognition, manipulation strategy, etc., in order to make the robot securely introduced in our daily support.

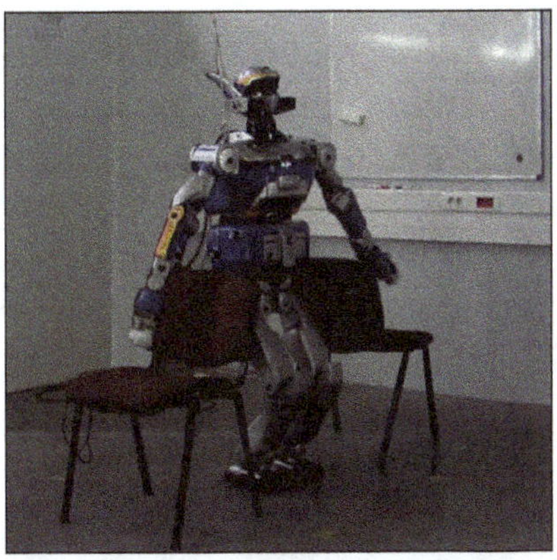

HRP2 in pushing work with its whole body motion.

Safety is also a crucial issue. Mental safety of humanoid causing no fear, no discomfort, no uneasiness, is psychologically evaluated as well as establishing physical safety.

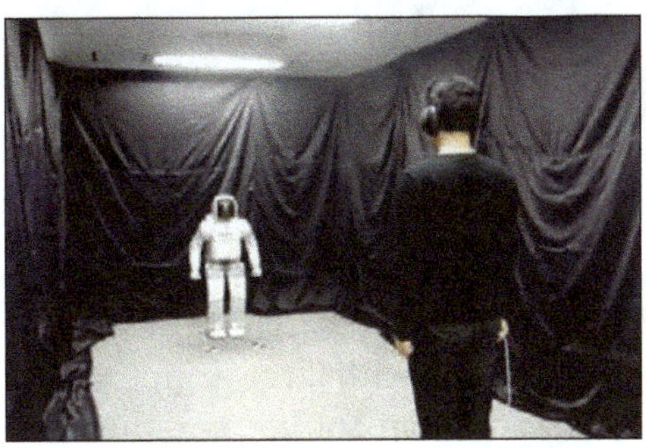

Mental safety evaluation with ASIMO.

Applications in Actual Construction Sites

Autonomous Wheel Loader

Autonomous wheel loader "Yamazumi-4" has succeeded in unmanned autonomous pile loading operation. The precise traveling control is required to repeat loading operations. Although actual wheel loader machines have the problem in controlling their precise traveling speed, the path planning and the path following control have been successfully designed by utilizing the odometer which can achieve precise geometric shape of the path.

Autonomous wheel loader.

Unmanned Construction

The unmanned construction is mostly required in the restoration of damaged local infrastructures as well as the prevention of further land erosions and damages to ensure workers safety after natural disasters. In 1990 large volcanic eruption at Unzen-Fugen-Dake brought significant damages to the surrounding area, and the unmanned construction methods have been applied for restoration works in upstream areas of the Mizunashi river where preparation of roads for construction, removal of soil and rock from earth accumulation basins, demolition of damaged factories, installation of blocks, construction of sand-trap dams, guide walls and flow control banks since 1994.

HRP in teleoperation task on construction machine.

Recently the unmanned transportation and the installation of a pre-cast arch culvert of about 15t in weight have been achieved. An arch was carried by the unmanned track to the site, and the unmanned shovel car with large gripper picked up the arch and fixed it in the pre-planned position with manual bolting, and then finally the unmanned survey was accomplished. It would be worthwhile to point out that humanoid may be applied in commercial machines to be transformed into remote-controlled construction machines as shown in figure.

Robotic House

Intelligent window system.

One of the most promising markets is robotic house, or house automation with robotic components. Current house has come to be equipped with automatically controlled actuators such as auto-shatter, auto window as well as remote controlled electric appliances. These devices can be regarded as a sort of Robotic Technology (RT) part. It will be beneficial if they are connected to a network, since users can obtain more services with RT collaborated motions and functions. A standard network device is crucial in order to achieve such a robotic house. The New Energy and Industrial Technology Development Organization (NEDO) has promoted the national robot project, "Open Innovation Promotion by Utilizing Basic Robotic Technology", that aims to develop common network modules for connecting RT parts in robotic house. A model room with intelligent window system has been designed and its prototype was built in AIST Tsukuba, to demonstrate security and energy saving air conditioning facility for robotic house.

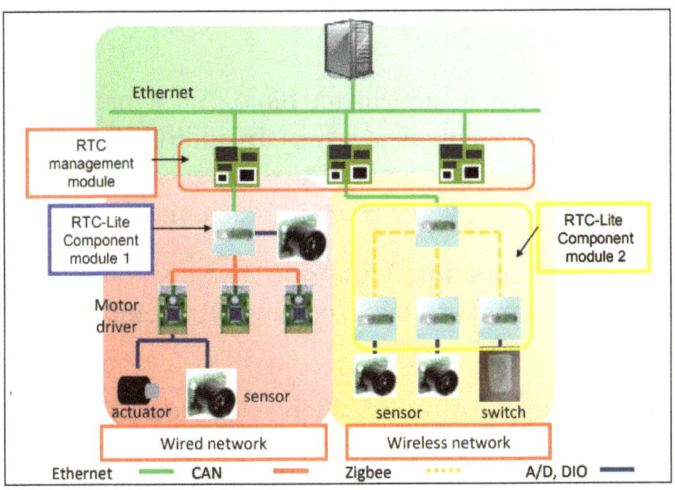

Network configuration and two type of RTCs.

Two types of RTCs have been proposed to facilitate smart and robust local network system in the robotic house. Power Line Communication (PLC) technology has been applied as basic infrastructure network in the robotic house. Based on two-layered network structure, robust home network system has been realized.

For the feasibility of achievement on this project, several RT products have been released as a practical trial in actual use. Area type human detection sensor by Kyokko Electronics, and motor driver unit by Okatech are first application products. Long term operation test has been accomplished to evaluate the robustness of developed network and RTCs framework.

(a) Motor driver. (b) Human detector.
Examples of device with mini RTC.

Medical Mechatronics

Mechatronics, the fusion of computer science with electrical and mechanical engineering, is used in numerous industries, but have you ever wondered why it has been so successful in the healthcare field?

Sensors can now detect physical, chemical and biological signals, which in turn can lead to earlier detection of a patient's medical issues. In addition, tiny mechatronic designs enable the creation of tools for endoscopic surgery, home monitoring and telemedicine, implanted devices, robotic surgical devices, and much more.

Thanks to medical mechatronics, numerous devices we all take for granted have been created such as; cardiac pacemakers, defibrillators, ultrasonic probes and arthroscopic devices.

Following are some of the applications of Medical Mechatronics:

- Prosthetic limbs that mimic realistic movements: To operate realistically, a prosthetic hand needs a separate motor for each finger, telling it how and when to move. One prosthetic hand, called bebionic has been created with microprocessors enabling precise manipulation. It can handle delicate movements like picking up an egg or threading a needle. Bebionic uses a combination of cutting-edge actuators, sensors, micro-processors, and embedded control software. The result is a prosthetic hand that mimics a real hand. LUKE arm has been approved by the FDA and is considered the "first device to be capable of making multiple and simultaneous movements through muscle contractions detected by electromyogram (EMG) electrodes." The arm has an onboard computer and built-in switches and motion sensors to execute movements.

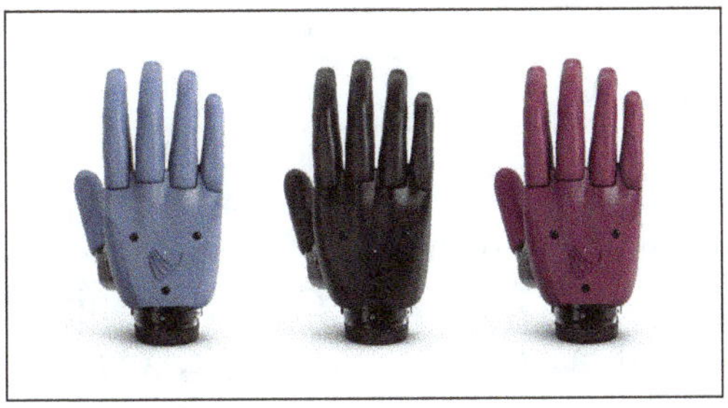

- Robotic-assisted surgical systems lead to quicker recovery: Surgeons around the world use a robotic-assisted, minimally invasive surgery platform called the da Vinci Surgical System. After a tiny incision is made on the patient, the surgeon manipulates the actions of the robotic arms. Mechatronics enable the robotic arms to move in sync with the surgeon's hand movements.
- Medical mechatronics in the future: Since medical mechatronics integrates many different domains—electric and mechanical system design, machine/deep learning algorithms, real-time clinical data analysis and assistive/rehabilitation robot development—it's a field that is sure to create many more life-saving tools for both patients and clinicians.

References

- Mechatronics-definition: mcgs.ch, Retrieved 14 January, 2020
- Mechatronics, definition: whatis.techtarget.com, Retrieved 08 July, 2020
- Mechatronic-system, topics-engineering: sciencedirect.com, Retrieved 29 March, 2020
- Integrated-mechatronic-design-for-servo-mechanical-systems, advances-in-mechatronics: intechopen.com, Retrieved 18 August, 2020
- Medical-mechatronics: captechu.edu, Retrieved 10 February, 2020

2

Digital Electronics

Digital electronics are referred to the types of electronics that operate on digital signals. It includes the concepts of number system, Boolean algebra, logic gates like AND, NOT, NAND, NOR, XOR and XNOR, combinatorial logic, sequential logic in synchronous and asynchronous circuit, etc. This chapter has been carefully written to provide an easy understanding of digital electronics.

Digital Electronics is the sub-branch of electronics which deals with digital signals for processing and controlling various systems and sub-systems. In various applications like sensors and actuators, usage of digital electronics is increasing extensively.

Digital Signals

Digital electronics is entirely the field in which digital signals is used. Digital signals are discretization of analog signals. A signal carries information. In digital signals the values in a particular band is same i.e. constant. Digital signals form the basis of digital circuit and digital electronics.

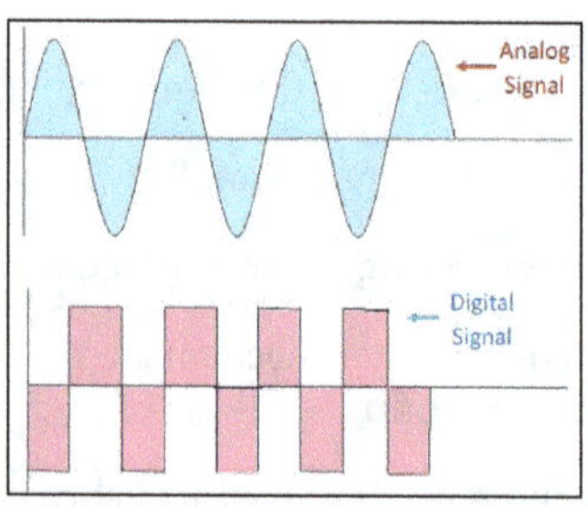

It is evident from the above diagram that analog signal is a continuous tome varying signal while digital signal consists of discrete values of voltage and thus it do not varies continuously with time.

Advantages of Digital Systems

- Easier designing: The Digital systems can be easily designed as they involve digital signals. These signals do not require exact value at a particular time but it consists of range of particular values of voltage. Thus, it comprises of basically two values 0 and 1 i.e. high or low.

- Noise immune: Digital systems are noise immune because digital signal consists of range of particular values. Thus, when noise is introduced in the medium and digital signal and analog signal both passes through it. The analog signal will be affected more because it varies continuous with time so it is difficult to identify that noise has destroyed which value of voltage. While in case of digital system, noise affect the particular range of the signal thus, it is clear to identify the particular range of filtering is also easy in case of digital signals.
- Information storage is simpler: The storage of information in digital systems is easy. It can be stored by latching thus; it can be stored for a long period of time.
- High accuracy and precision: The digital signal offers high accuracy and precision. This is because the processing of digital signal is done through the switching circuit. While in case of analog signals the processing and its output is highly dependent on circuit components. The accuracy obtained in analog circuits is restricted to 3-4 digits while in case of digital signals the accuracy is far more than the analog circuit.
- Programmable: The digital systems are easily programmable but analog system becomes complex when excessive programming of components is done.

Disadvantages of Digital Systems

Everything comes up with pros and cons. Similarly digital systems too hold certain disadvantages which are as follows:

- Expensive: Digital systems are expensive because it involves switching elements.
- Analog nature of real world entities: We need to convert the digital output in analog form because all the real world entities are analog.

Number System

In digital electronics, the number system is used for representing the information. The number system has different bases and the most common of them are the decimal, binary, octal, and hexadecimal. The base or radix of the number system is the total number of the digit used in the number system. Suppose if the number system representing the digit from 0 – 9 then the base of the system is the 10.

Types of Number Systems

Some of the important types of number system are:

- Decimal Number System.

- Binary Number System.
- Octal Number System.
- Hexadecimal Number System.

These number systems are explained below in details:

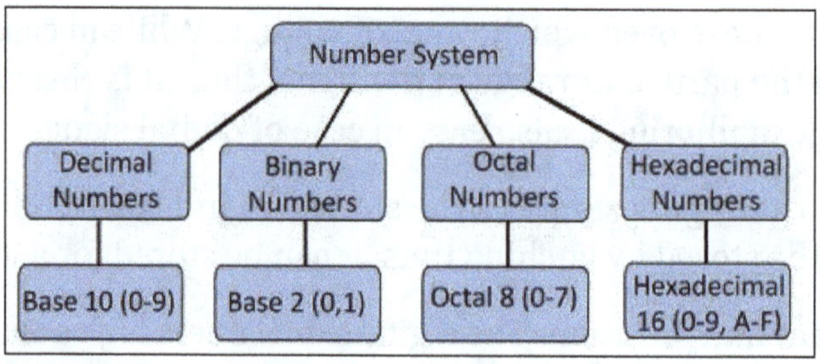

Decimal Number Systems

The number system is having digit 0, 1, 2, 3, 4, 5, 6, 7, 8, 9; this number system is known as a decimal number system because total ten digits are involved. The base of the decimal number system is 10.

Binary Number Systems

The modern computers do not process decimal number; they work with another number system known as a binary number system which uses only two digits 0 and 1. The base of binary number system is 2 because it has only two digits 0 and 1. The digital electronic equipment's are works on the binary number system and hence the decimal number system is converted into binary system.

The table is shown below the decimal, binary, octal, and hexadecimal numbers from 0 to 15 and their equivalent binary number.

Decimal	Binary	Octal	Hexadecimal
0	0000	0	0
1	0001	1	1
2	0010	2	2
3	0011	3	3
4	0100	4	4
5	0101	5	5
6	0110	6	6
7	0111	7	7
8	1000	10	8
9	1001	11	9
10	1010	12	A
11	1011	13	B

12	1100	14	C
13	1101	15	D
14	1110	16	E
15	1111	17	F

Octal Numbers

The base of a number system is equal to the number of digits used, i.e., for decimal number system the base is ten while for the binary system the base is two. The octal system has the base of eight as it uses eight digits 0, 1, 2, 3, 4, 5, 6, 7.

All these digits from 0 to 7 have the same physical meaning as by decimal symbols, the next digit in the octal number is represented by 10, 11, 12, which are equivalent to decimal digits 8, 9, 10 respectively. In this way, the octal number 20 will represent the decimal digit and subsequently, 21, 22, 23. Octal numbers will represent the decimal number digit 17, 18, 19... etc. and so on.

Hexadecimal Numbers

These numbers are used extensively in microprocessor work. The hexadecimal number system has a base of 16, and hence it consists of the following sixteen numbers of digits.

0, 1, 2, 3, 4, 5, 6, 7, 8, 9, A, B, C, D, E, F.

The size of the hexadecimal is much shorter than the binary number which makes them easy to write and remember. Let 0000 to 000F representing hexadecimal numbers from zero to fifteen, then 0010, 0011, 0012,...etc., will represent sixteen, seventeen, eighteen...etc. till 001F which represent thirty open and so on.

Boolean Algebra

Boolean Algebra is the mathematics we use to analyse digital gates and circuits. We can use these "Laws of Boolean" to both reduce and simplify a complex Boolean expression in an attempt to reduce the number of logic gates required. Boolean Algebra is therefore a system of mathematics based on logic that has its own set of rules or laws which are used to define and reduce Boolean expressions.

The variables used in Boolean Algebra only have one of two possible values, a logic "0" and a logic "1" but an expression can have an infinite number of variables all labelled individually to represent inputs to the expression, For example, variables A, B, C etc, giving us a logical expression of A + B = C, but each variable can ONLY be a 0 or a 1. Examples of these individual laws of Boolean, rules and theorems for Boolean Algebra are given in the table.

Table: Truth Tables for the Laws of Boolean.

Boolean Expression	Description	Equivalent Switching Circuit	Boolean Algebra Law or Rule
$A + 1 = 1$	A in parallel with closed = "CLOSED"	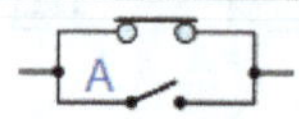	Annulment
$A + 0 = A$	A in parallel with open = "A"	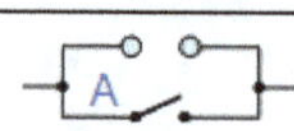	Identity
$A . 1 = A$	A in series with closed = "A"	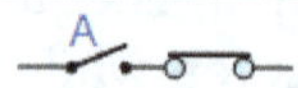	Identity
$A . 0 = 0$	A in series with open = "OPEN"		Annulment
$A + A = A$	A in parallel with A = "A"	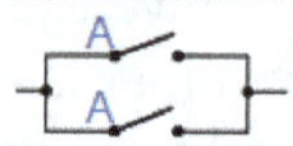	Idempotent
$A . A = A$	A in series with A = "A"	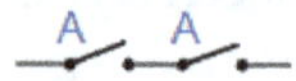	Idempotent
NOT $\overline{A} = A =$	NOT NOT A (double negative) = "A"		Double Negation
$A + \overline{A} = 1$	A in parallel with NOT A = "CLOSED"	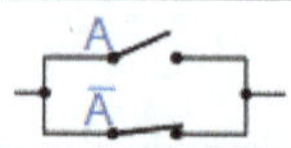	Complement
$A . \overline{A} = 0$	A in series with NOT A = "OPEN"	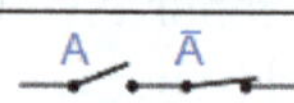	Complement
$A+B = B+A$	A in parallel with B = B in parallel with A	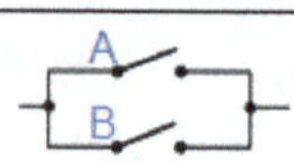	Commutative
$A.B = B.A$	A in series with B = B in series with A	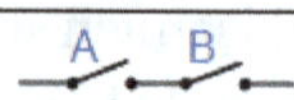	Commutative
$\overline{A+B} = \overline{A}.\overline{B}$	Invert and replace OR with AND		de Morgan's Theorem
$\overline{A}.\overline{B} = \overline{A+B}$	Invert and replace AND with OR		de Morgan's Theorem

The basic Laws of Boolean Algebra that relate to the Commutative Law allowing a change in position for addition and multiplication, the Associative Law allowing the removal of brackets for addition and multiplication, as well as the Distributive Law allowing the factoring of an expression, are the same as in ordinary algebra.

Each of the Boolean Laws above are given with just a single or two variables, but the number of variables defined by a single law is not limited to this as there can be an infinite number of variables as inputs too the expression. These Boolean laws can be used to prove any given Boolean expression as well as for simplifying complicated digital circuits.

Annulment Law

A term AND´ed with a "0" equals 0 or OR´ed with a "1" will equal 1.

- $A . 0 = 0$: A variable AND'ed with 0 is always equal to 0.
- $A + 1 = 1$: A variable OR'ed with 1 is always equal to 1.

Identity Law

A term OR´ed with a "0" or AND´ed with a "1" will always equal that term.

- $A + 0 = A$: A variable OR'ed with 0 is always equal to the variable.
- $A \cdot 1 = A$: A variable AND'ed with 1 is always equal to the variable.

Idempotent Law

An input that is AND´ed or OR´ed with itself is equal to that input.

- $A + A = A$: A variable OR'ed with itself is always equal to the variable.
- $A \cdot A = A$: A variable AND'ed with itself is always equal to the variable.

Complement Law

A term AND´ed with its complement equals "0" and a term OR´ed with its complement equals "1".

- $A.\overline{A} = 0$: A variable AND'ed with its complement is always equal to 0.
- $A + \overline{A} = 1$: A variable OR'ed with its complement is always equal to 1.

Commutative Law

The order of application of two separate terms is not important:

- $A \cdot B = B \cdot A$: The order in which two variables are AND'ed makes no difference.
- $A + B = B + A$: The order in which two variables are OR'ed makes no difference.

Double Negation Law

A term that is inverted twice is equal to the original term.

- $A = A$: A double complement of a variable is always equal to the variable.

de Morgan´s Theorem

There are two "de Morgan´s" rules or theorems:

- Two separate terms NOR´ed together is the same as the two terms inverted (Complement) and AND´ed for example: $\overline{A+B} = \overline{A} + \overline{B}$.
- Two separate terms NAND´ed together is the same as the two terms inverted (Complement) and OR´ed for example: $\overline{A.B} = \overline{A} + \overline{B}$.

Other algebraic Laws of Boolean include:

Distributive Law

This law permits the multiplying or factoring out of an expression.

- A(B + C) = A.B + A.C (OR Distributive Law).
- A + (B.C) = (A + B).(A + C) (AND Distributive Law).

Absorptive Law

This law enables a reduction in a complicated expression to a simpler one by absorbing like terms.

- A + (A.B) = A (OR Absorption Law).
- A(A + B) = A (AND Absorption Law).

Associative Law

This law allows the removal of brackets from an expression and regrouping of the variables.

- A + (B + C) = (A + B) + C = A + B + C (OR Associate Law).
- A(B.C) = (A.B)C = A . B . C (AND Associate Law).

Boolean Algebra Functions

Using the information above, simple 2-input AND, OR and NOT Gates can be represented by 16 possible functions as shown in the following table:

Function	Description	Expression
1.	NULL	0
2.	IDENTITY	1
3.	Input A	A
4.	Input B	B
5.	NOT A	$\overline{A}$
6.	NOT B	$\overline{B}$
7.	A AND B (AND)	A . B
8.	A AND NOT B	$A.\overline{B}$
9.	NOT A AND B	$\overline{A}.B$
10.	NOT AND (NAND)	$\overline{A.B}$
11.	A OR B (OR)	A + B

12.	A OR NOT B	$A.\overline{B}$
13.	NOT A OR B	$\overline{A}+B$
14.	NOT OR (NOR)	$\overline{A+B}$
15.	Exclusive-OR	$A.\overline{B}+\overline{A}.B$
16.	Exclusive-NOR	$A.B+\overline{A}.\overline{B}$

Laws of Boolean Algebra Example

Using the above laws, simplify the following expression: (A + B)(A + C).

Q = (A + B).(A + C)	
A.A + A.C + A.B + B.C	– Distributive law
A + A.C + A.B + B.C	– Idempotent AND law (A.A = A)
A(1 + C) + A.B + B.C	– Distributive law
A.1 + A.B + B.C	– Identity OR law (1 + C = 1)
A(1 + B) + B.C	– Distributive law
A.1 + B.C	– Identity OR law (1 + B = 1)
Q = A + (B.C)	– Identity AND law (A.1 = A)

Then the expression: (A + B)(A + C) can be simplified to A + (B.C) as in the Distributive law.

Logic Gates

A logic gate is a building block of a digital circuit. Most logic gates have two inputs and one output and are based on Boolean algebra. At any given moment, every terminal is in one of the two binary conditions false (high) or true (low). False represents 0, and true represents 1. Depending on the type of logic gate being used and the combination of inputs, the binary output will differ. A logic gate can be thought of like a light switch, wherein one position the output is off—0, and in another, it is on—1. Logic gates are commonly used in integrated circuits (IC).

AND Gate

The AND gate is a basic digital logic gate that implements logical conjunction - it behaves according to the truth table to the right. A HIGH output (1) results only if all the inputs to the AND gate are HIGH (1). If none or not all inputs to the AND gate are HIGH, a LOW output results. The function can be extended to any number of inputs.

INPUT		OUTPUT
A	B	A AND B
0	0	0
0	1	0
1	0	0
1	1	1

Symbols

There are three symbols for AND gates: the American (ANSI or 'military') symbol and the IEC ('European' or 'rectangular') symbol, as well as the deprecated DIN symbol. Additional inputs can be added as needed.

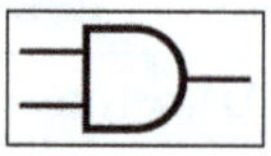
MIL/ANSI Symbol.

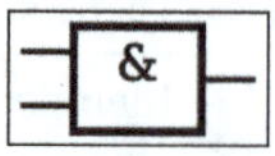

IEC Symbol.

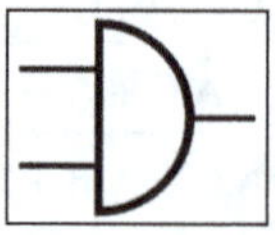
DIN Symbol.

The AND gate with inputs A and B and output C implements the logical expression $C = A \cdot B$. This expression also may be denoted as C=A^B or C=A&B.

Implementations

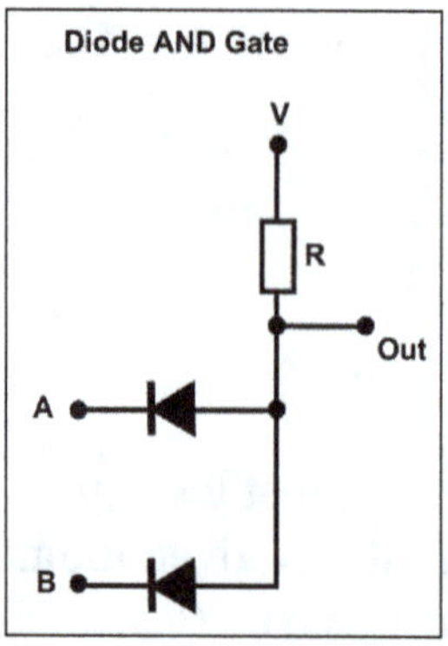

AND gate using diodes.

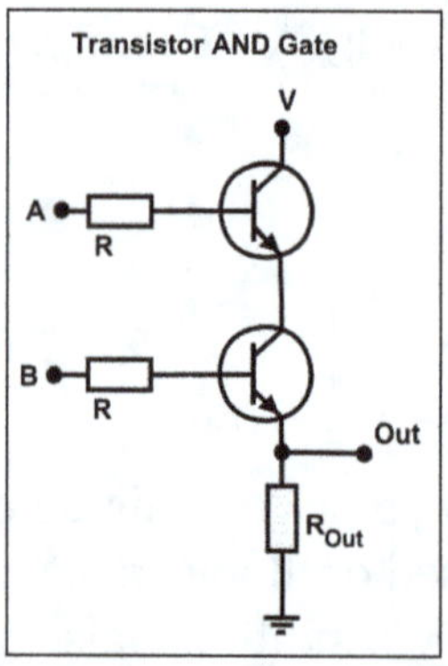

AND gate using transistors.

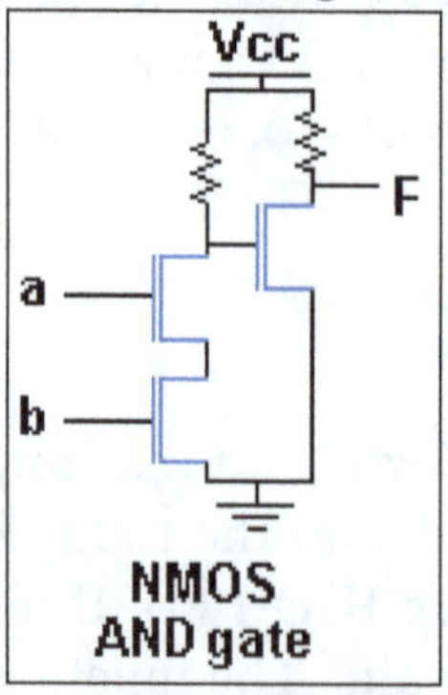

NMOS AND gate.

An AND gate is usually designed using N-channel (pictured) or P-channel MOSFETs. The digital inputs **a** and **b** cause the output **F** to have the same result as the AND function.

Analytical Representation

$f(a,b) = a * b$ is the analytical representation of AND gate:

- $f(0,0) = 0*0 = 0$
- $f(0,1) = 0*1 = 0$
- $f(1,0) = 1*0 = 0$
- $f(1,1) = 1*1 = 1$

Alternatives

If no specific AND gates are available, one can be made from NAND or NOR gates, because NAND and NOR gates are considered the "universal gates," meaning that they can be used to make all the others.

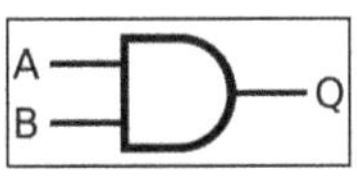

Desired gate.

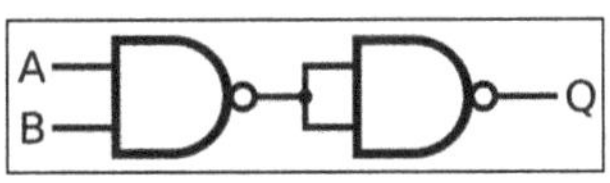

NAND construction.

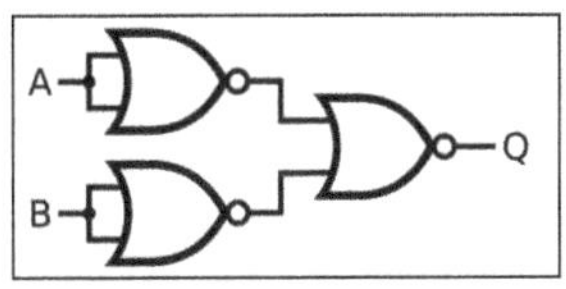

NOR construction.

IC Package

AND gates are available in IC packages. The 7408 IC is a well-known QUAD 2-Input AND GATES and contains four independent gates each of which performs the logic AND function.

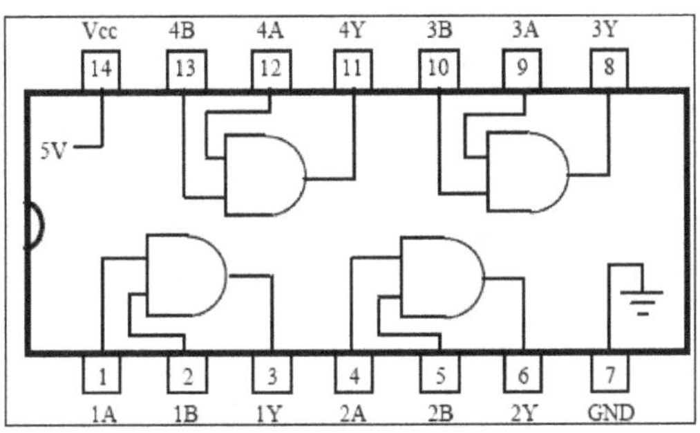

IC 7408.

OR Gate

The OR gate is a digital logic gate that implements logical disjunction – it behaves according to the truth table to the right. A HIGH output (1) results if one or both the inputs to the gate are HIGH (1). If neither input is high, a LOW output (0) results. In another sense, the function of OR effectively finds the maximum between two binary digits, just as the complementary AND function finds the minimum.

INPUT		OUTPUT
A	B	A OR B
0	0	0
0	1	1
1	0	1
1	1	1

Symbols

There are two symbols of OR gates: the American (ANSI or 'military') symbol and the IEC ('European' or 'rectangular') symbol, as well as the deprecated DIN symbol.

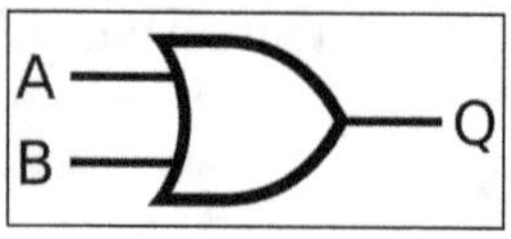

MIL/ANSI Symbol.

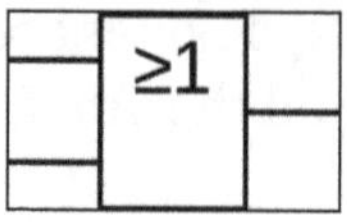

IEC Symbol.

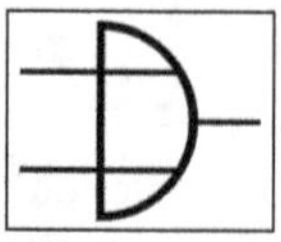

DIN Symbol.

Hardware Description and Pinout

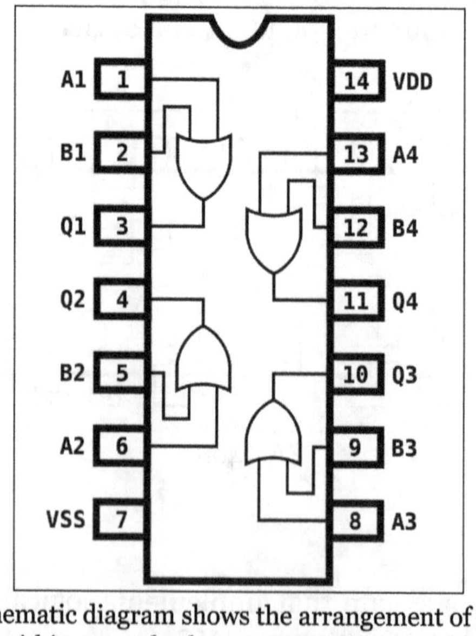

This schematic diagram shows the arrangement of four OR gates within a standard 4071 CMOS integrated circuit.

OR gates are basic logic gates, and are available in TTL and CMOS ICs logic families. The standard 4000 series CMOS IC is the 4071, which includes four independent two-input OR gates. The TTL device is the 7432. There are many offshoots of the original 7432 OR gate, all having the same pinout but different internal architecture, allowing them to operate in different voltage ranges and/or at higher speeds. In addition to the standard 2-input OR gate, 3- and 4-input OR gates are also available. In the CMOS series, these are:

- 4075: Triple 3-input OR gate.
- 4072: Dual 4-input OR gate.

Variations include:

- 74LS32: Quad 2-input OR gate (low power Schottky version).
- 74HC32: Quad 2-input OR gate (high speed CMOS version) - has lower current consumption/wider voltage range.
- 74AC32: Quad 2-input OR gate (advanced CMOS version) - similar to 74HC32, but with significantly faster switching speeds and stronger drive.
- 74LVC32: Low voltage CMOS version of the same.

Implementations

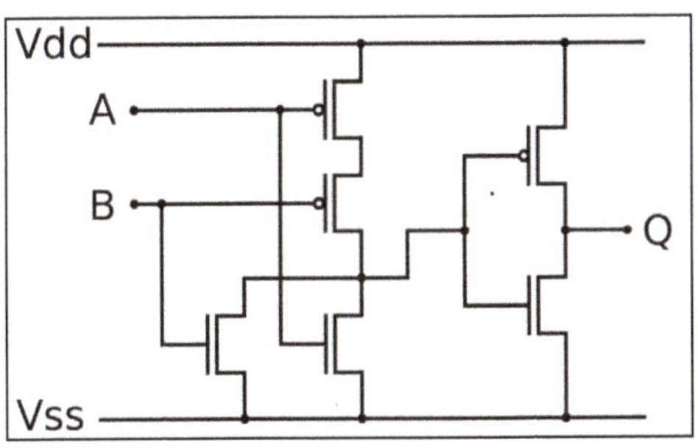

CMOS OR gate.

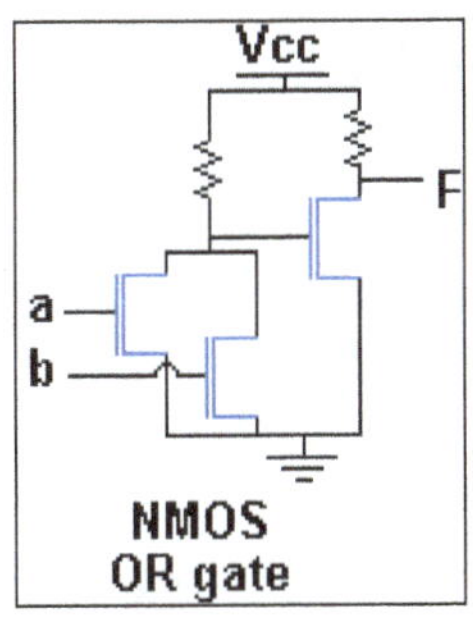

NMOS OR gate.

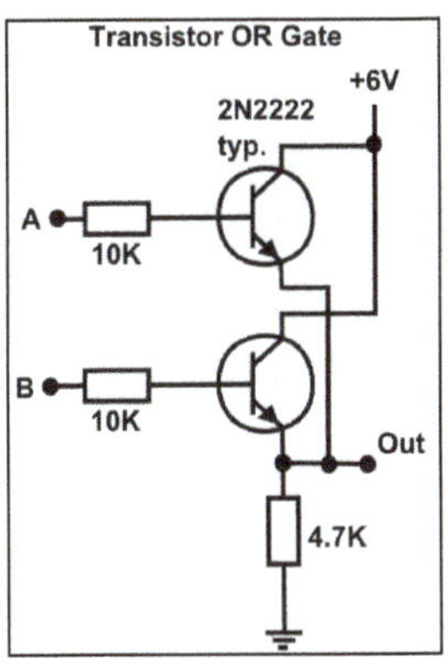

BJT OR gate.

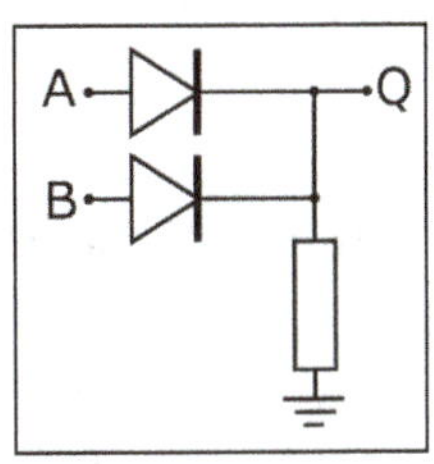

OR gate using diodes.

Analytical Representation

$f(a,b) = a + b - a * b$ is the analytical representation of OR gate:

- $f(0,0) = 0 + 0 - 0 * 0 = 0$
- $f(0,1) = 0 + 1 - 0 * 1 = 1$
- $f(1,0) = 1 + 0 - 1 * 0 = 1$
- $f(1,1) = 1 + 1 - 1 * 1 = 1$

Alternatives

If no specific OR gates are available, one can be made from NAND or NOR gates in the configuration shown in the image below. Any logic gate can be made from a combination of NAND or NOR gates.

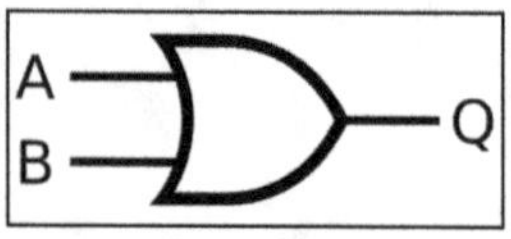

Desired gate.

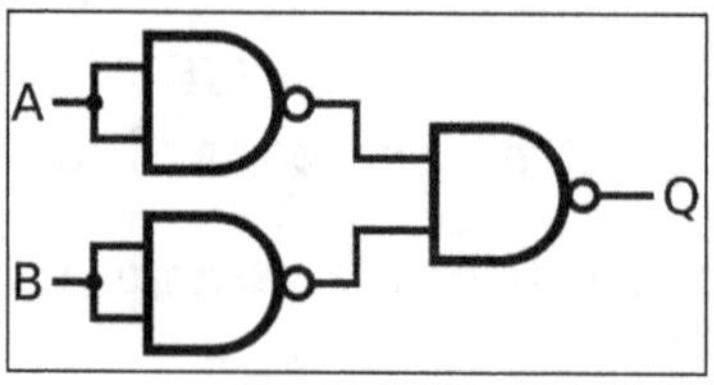

NAND construction.

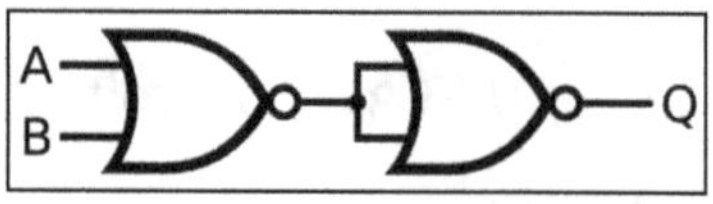

NOR construction.

Wired-OR

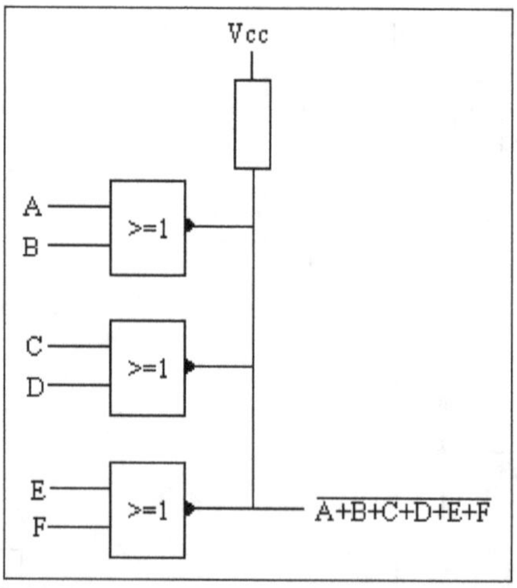

Wired OR gate using open-collector NOR gates.

With active low open collector logic outputs, as used for control signals in many circuits, an OR function can be produced by wiring together several outputs. This arrangement is called a *wired OR*. This implementation of an OR function typically is also found in integrated circuits of N or P-type only transistor processes.

NOT Gate

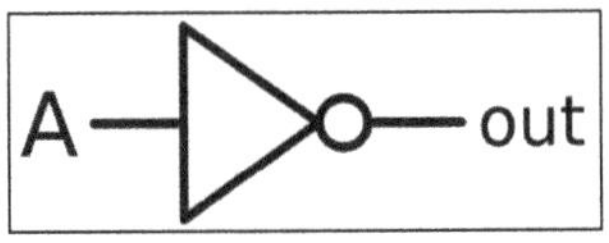

Traditional NOT Gate (Inverter) symbol.

In digital logic, an inverter or NOT gate is a logic gate which implements logical negation. The truth table is shown below:

INPUT	OUTPUT
A	NOT A
0	1
1	0

Electronic Implementation

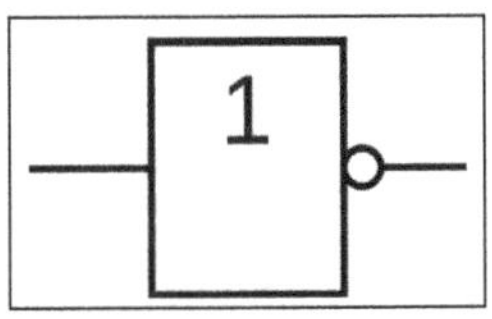

International Electrotechnical Commission
NOT Gate (Inverter) symbol.

An inverter circuit outputs a voltage representing the opposite logic-level to its input. Its main function is to invert the input signal applied. If the applied input is low then the output becomes high and vice versa. Inverters can be constructed using a single NMOS transistor or a single PMOS transistor coupled with a resistor. Since this 'resistive-drain' approach uses only a single type of transistor, it can be fabricated at a low cost. However, because current flows through the resistor in one of the two states, the resistive-drain configuration is disadvantaged for power consumption and processing speed. Alternatively, inverters can be constructed using two complementary transistors in a CMOS configuration. This configuration greatly reduces power consumption since one of the transistors is always off in both logic states. Processing speed can also be improved due to the relatively low resistance compared to the NMOS-only or PMOS-only type devices. Inverters can also be constructed with bipolar junction transistors (BJT) in either a resistor–transistor logic (RTL) or a transistor–transistor logic (TTL) configuration.

Digital electronics circuits operate at fixed voltage levels corresponding to a logical 0 or 1. An inverter circuit serves as the basic logic gate to swap between those two voltage levels. Implementation determines the actual voltage, but common levels include (0, +5V) for TTL circuits.

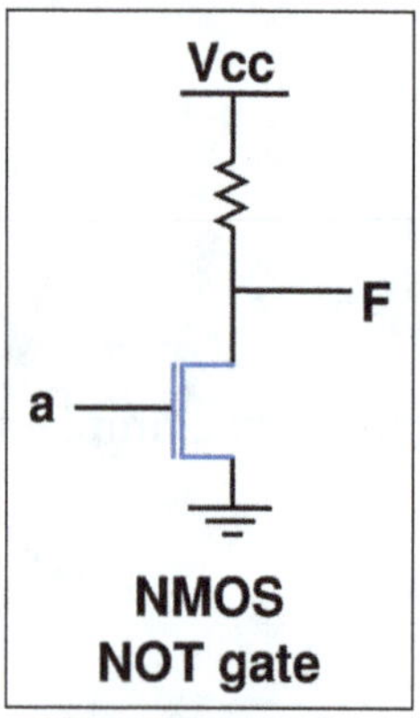

NMOS inverter.

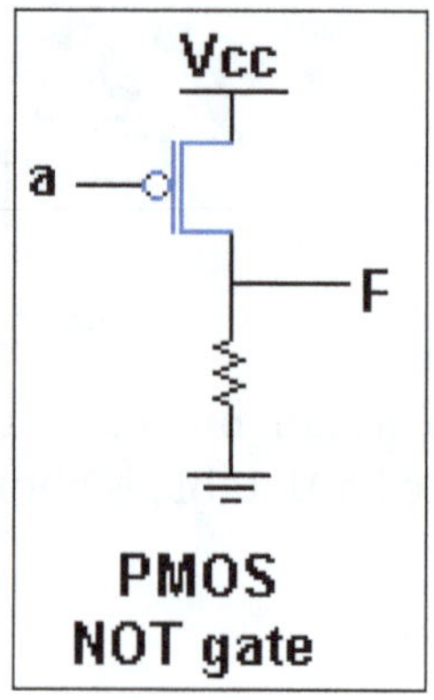

PMOS inverter.

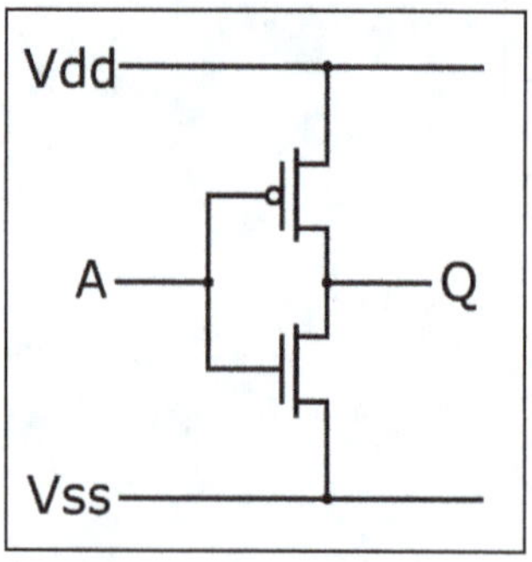

Static CMOS inverter.

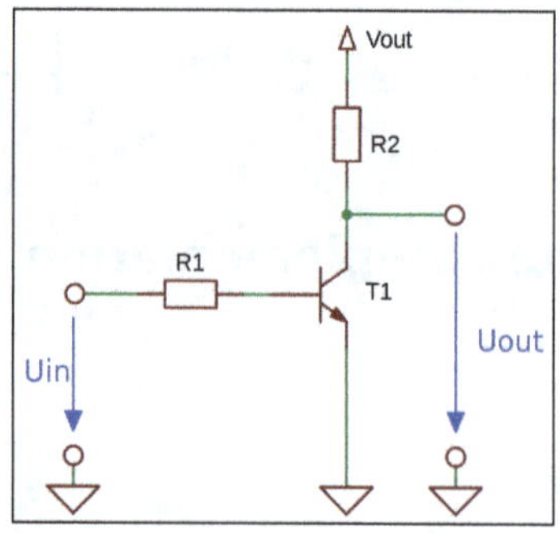

NPN resistor–transistor logic inverter.

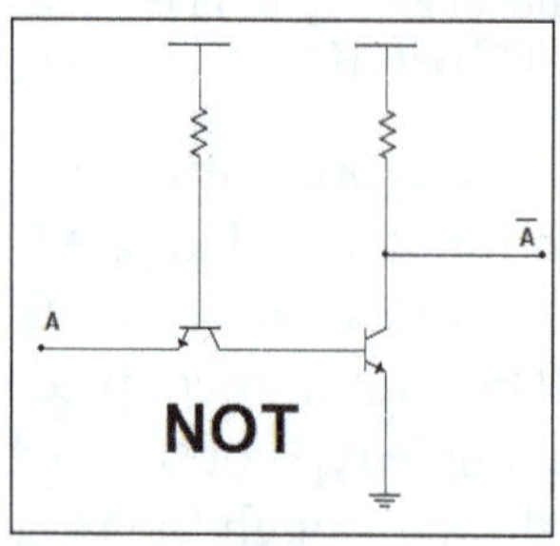

NPN transistor–transistor logic inverter.

Digital Building Block

This schematic diagram shows the arrangement of NOT gates within a standard 4049 CMOS hex inverting buffer.

The inverter is a basic building block in digital electronics. Multiplexers, decoders, state machines, and other sophisticated digital devices may use inverters.

The *hex inverter* is an integrated circuit that contains six (*hexa-*) inverters. For example, the 7404 TTL chip which has 14 pins and the 4049 CMOS chip which has 16 pins, 2 of which are used for power/referencing, and 12 of which are used by the inputs and outputs of the six inverters (the 4049 has 2 pins with no connection).

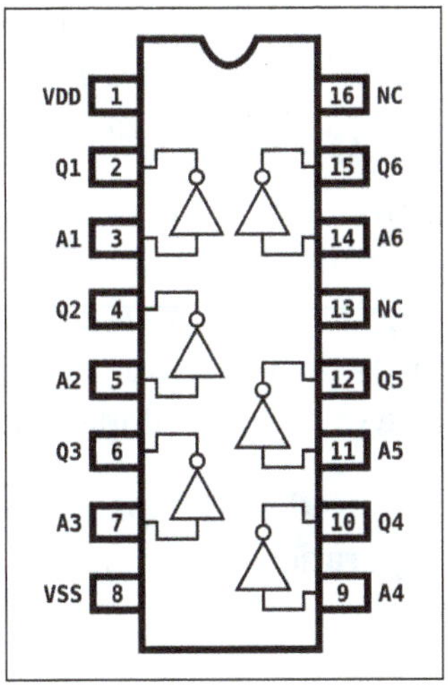

Analytical Representation

$f(a) = 1 - a$ is the analytical representation of NOT gate:

- $f(0) = 1 - 0 = 1$
- $f(1) = 1 - 1 = 0$

Alternatives

If no specific NOT gates are available, one can be made from the universal NAND or NOR gates.

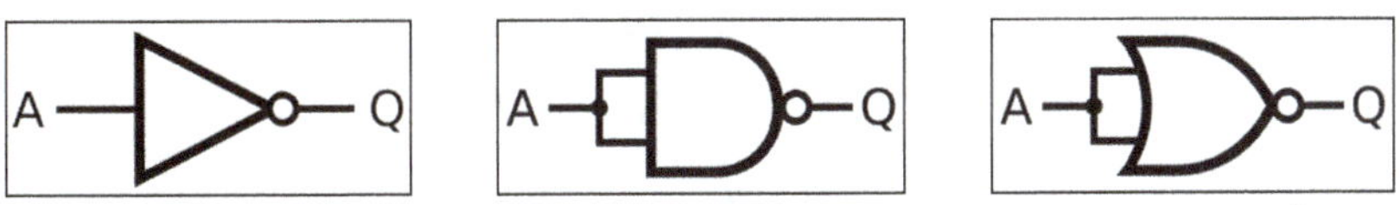

Desired gate. NAND construction. NOR construction.

Performance Measurement

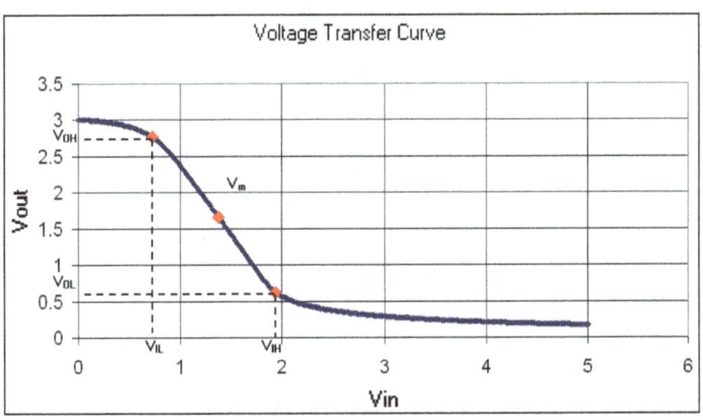

Voltage transfer curve for a 20 μm inverter constructed.

Digital inverter quality is often measured using the voltage transfer curve (VTC), which is a plot of output vs. input voltage. From such a graph, device parameters including noise tolerance, gain, and operating logic levels can be obtained.

Ideally, the VTC appears as an inverted step function – this would indicate precise switching between *on* and *off* – but in real devices, a gradual transition region exists. The VTC indicates that for low input voltage, the circuit outputs high voltage; for high input, the output tapers off towards the low level. The slope of this transition region is a measure of quality – steep (close to infinity) slopes yield precise switching.

The tolerance to noise can be measured by comparing the minimum input to the maximum output for each region of operation (on/off).

NAND Gate

In digital electronics, a NAND gate (NOT-AND) is a logic gate which produces an output which is false only if all its inputs are true; thus its output is complement to that of an AND gate. A LOW (0) output results only if all the inputs to the gate are HIGH (1); if any input is LOW (0), a HIGH (1) output results. A NAND gate is made using transistors and junction diodes. By De Morgan's theorem, a two-input NAND gate's logic may be expressed as $\overline{AB} = \overline{A} + \overline{B}$, making a NAND gate equivalent to inverters followed by an OR gate.

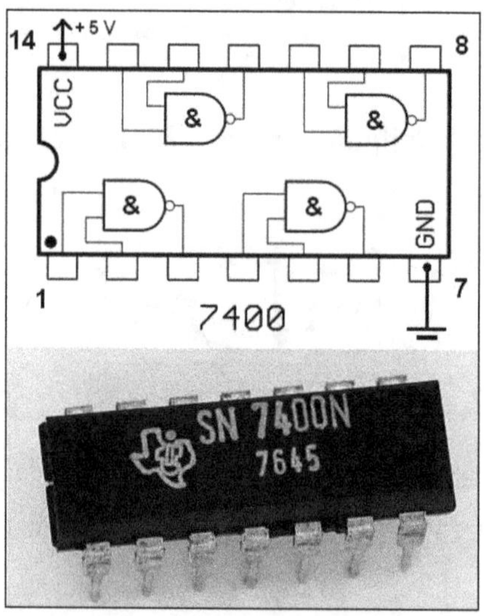

The TTL 7400 chip, containing four NANDs. The two additional pins supply power (+5 V) and connect the ground.

INPUT		OUTPUT
A	B	A NAND B
0	0	1
0	1	1
1	0	1
1	1	0

The NAND gate is significant because any boolean function can be implemented by using a combination of NAND gates. This property is called functional completeness. It shares this property with the NOR gate. Digital systems employing certain logic circuits take advantage of NAND's functional completeness.

The function NAND(a_1, a_2, ..., a_n) is logically equivalent to NOT(a_1 AND a_2 AND ... AND a_n).

One way of expressing A NAND B is $\overline{A \wedge B}$, where the symbol $\wedge$ signifies AND and the bar signifies the negation of the expression under it: in essence, simply $\neg(A \wedge B)$.

NAND gates with two or more inputs are available as integrated circuits in transistor-transistor logic, CMOS, and other logic families.

Symbols

There are three symbols for NAND gates: The MIL/ANSI symbol, the IEC symbol and the deprecated DIN symbol sometimes found on old schematics. The ANSI symbol for the NAND gate is a standard AND gate with an inversion bubble connected.

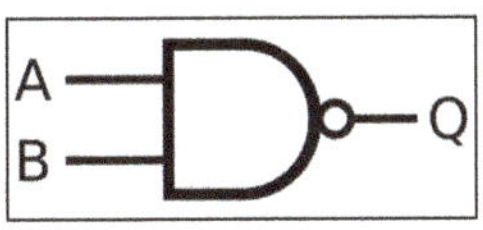

MIL/ANSI Symbol.

IEC Symbol.

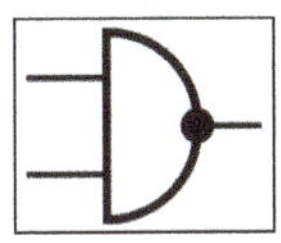

DIN Symbol.

Hardware Description and Pinout

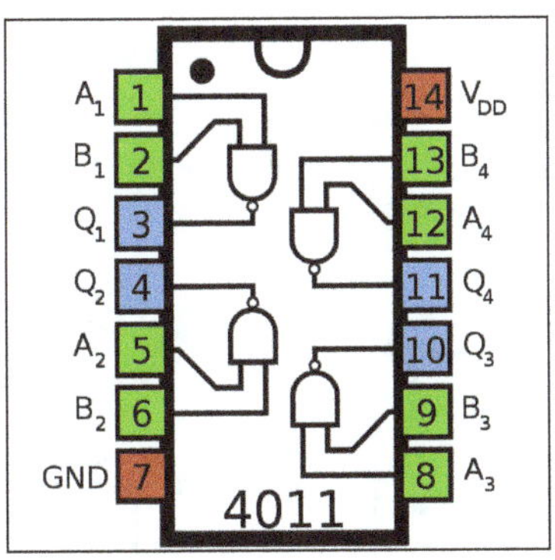

Diagram of the NAND gates in a CMOS type 4011 integrated circuit.

NAND gates are basic logic gates, and as such they are recognised in TTL and CMOS ICs.

CMOS Version

The standard, 4000 series, CMOS IC is the 4011, which includes four independent, two-input, NAND gates.

Availability

These devices are available from most semiconductor manufacturers such as Fairchild Semiconductor, Philips or Texas Instruments. These are usually available in both through-hole DIL and SOIC format. Datasheets are readily available in most datasheet databases.

The standard 2-, 3-, 4- and 8-input NAND gates are available:

- CMOS:
    - 4011: Quad 2-input NAND gate.
    - 4023: Triple 3-input NAND gate.
    - 4012: Dual 4-input NAND gate.
    - 4068: Mono 8-input NAND gate.
- TTL:
    - 7400: Quad 2-input NAND gate.
    - 7410: Triple 3-input NAND gate.
    - 7420: Dual 4-input NAND gate.
    - 7430: Mono 8-input NAND gate.

Implementations

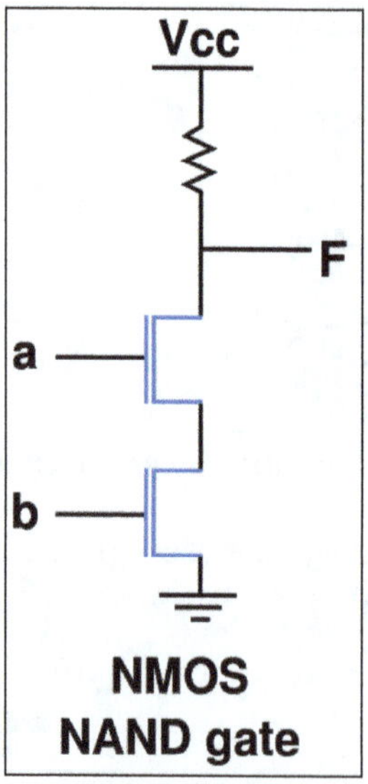

PMOS NAND gate.

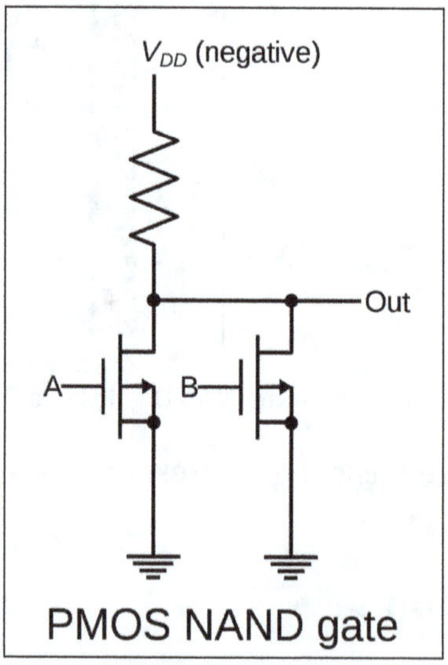

PMOS NAND gate.

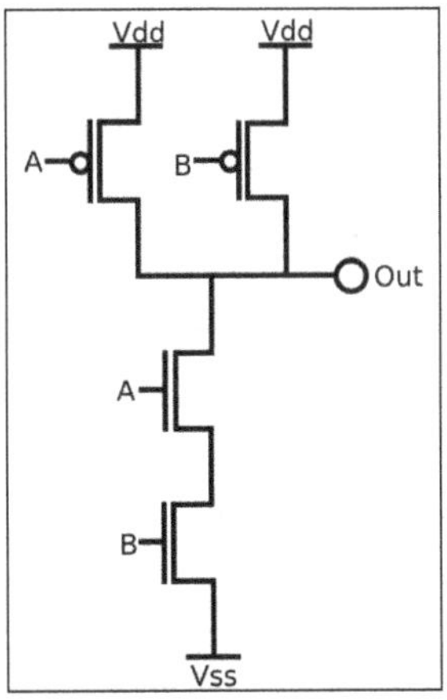

CMOS NAND gate.

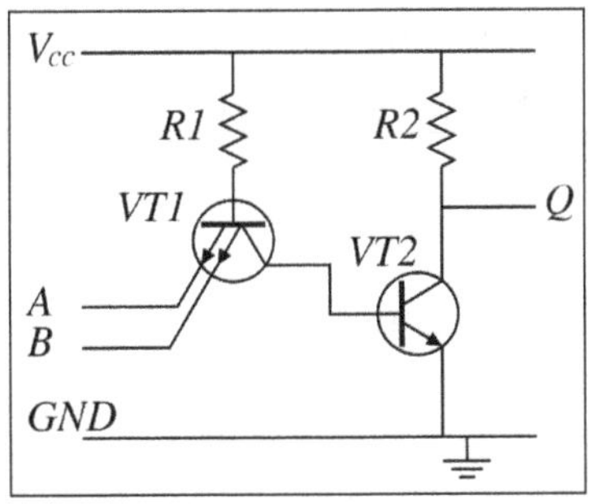

TTL NAND gate.

Functional Completeness

The NAND gate has the property of functional completeness, which it shares with the NOR gate. That is, any other logic function (AND, OR, etc.) can be implemented using only NAND gates. An entire processor can be created using NAND gates alone. In TTL ICs using multiple-emitter transistors, it also requires fewer transistors than a NOR gate.

As NOR gates are also functionally complete, if no specific NAND gates are available, one can be made from NOR gates.

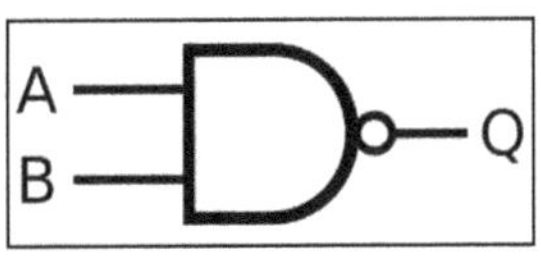

Desired gate.

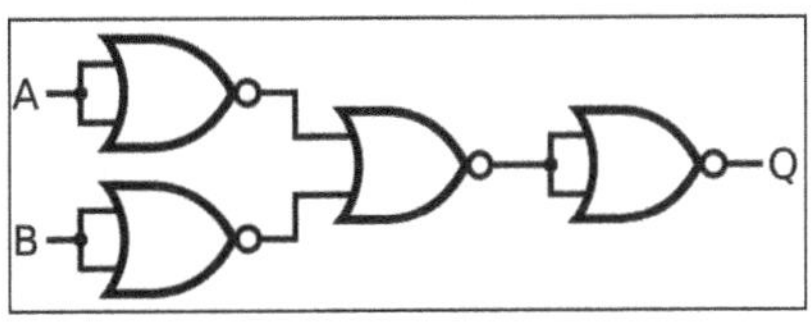

NOR Construction.

NOR Gate

The NOR gate is a digital logic gate that implements logical NOR - it behaves according to the truth table to the right. A HIGH output (1) results if both the inputs to the gate are LOW (0); if one or both input is HIGH (1), a LOW output (0) results. NOR is the result of the negation of the OR operator. It can also in some senses be seen as the inverse of an AND gate. NOR is a functionally complete operation—NOR gates can be combined to generate any other logical function. It shares this property with the NAND gate. By contrast, the OR operator is monotonic as it can only change LOW to HIGH but not vice versa.

INPUT		OUTPUT
A	B	A NOR B
0	0	1

0	1	0
1	0	0
1	1	0

In most, but not all, circuit implementations, the negation comes for free—including CMOS and TTL. In such logic families, OR is the more complicated operation; it may use a NOR followed by a NOT. A significant exception is some forms of the domino logic family.

The original Apollo Guidance Computer used 4,100 integrated circuits (IC), each one containing only two 3-input NOR gates.

Symbols

There are three symbols for NOR gates: The American (ANSI or 'military') symbol and the IEC ('European' or 'rectangular') symbol, as well as the deprecated DIN symbol. The ANSI symbol for the NOR gate is a standard OR gate with an inversion bubble connected.

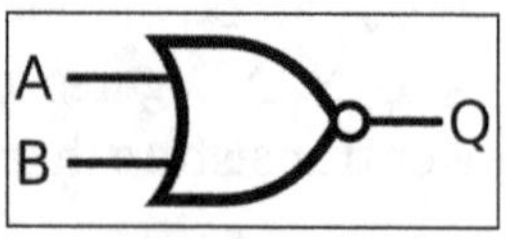

MIL/ANSI Symbol.

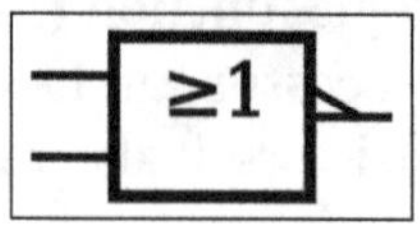

IEC Symbol.

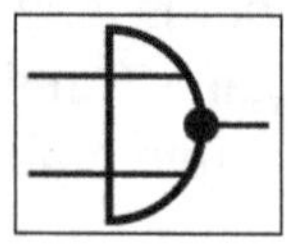

DIN Symbol.

Hardware Description and Pinout

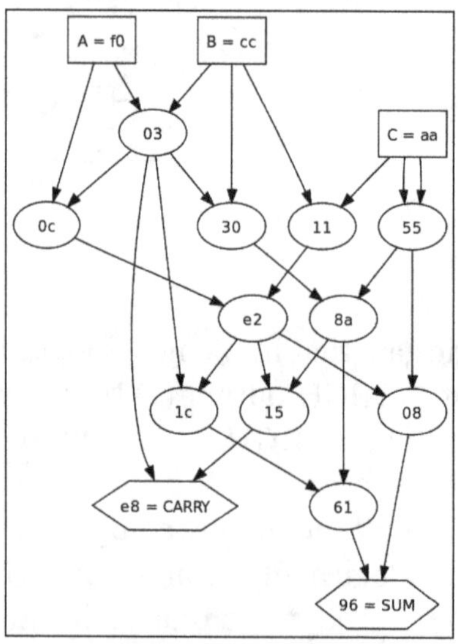

NOR Full adder.

NOR Gates are basic logic gates, and as such they are recognised in TTL and CMOS ICs. The standard, 4000 series, CMOS IC is the 4001, which includes four independent, two-input, NOR gates.

The pinout diagram is as follows:

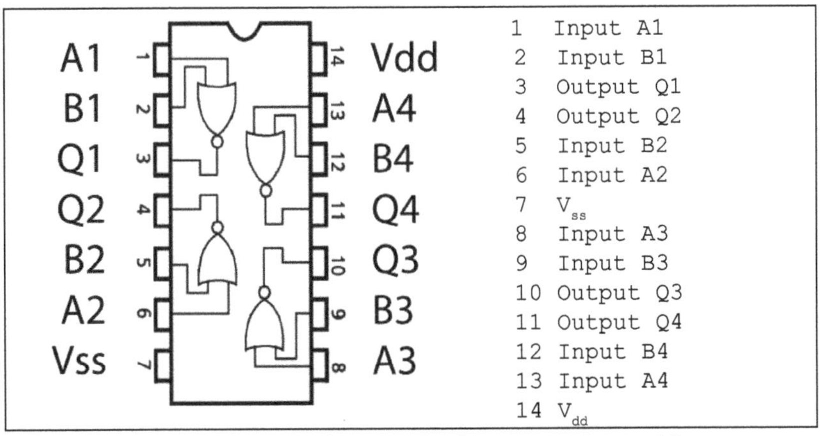

Pinout Diagram of a 4001 Quad NOR DIP-format IC.

Availability

These devices are available from most semiconductor manufacturers such as Fairchild Semiconductor, Philips or Texas Instruments. These are usually available in both through-hole DIP and SOIC format. Datasheets are readily available in most datasheet databases.

In the popular CMOS and TTL logic families, NOR gates with up to 8 inputs are available:

- CMOS:
    - 4001: Quad 2-input NOR gate.
    - 4025: Triple 3-input NOR gate.
    - 4002: Dual 4-input NOR gate.
    - 4078: Single 8-input NOR gate.
- TTL:
    - 7402: Quad 2-input NOR gate.
    - 7427: Triple 3-input NOR gate.
    - 7425: Dual 4-input NOR gate (with strobe, obsolete).
    - 74260: Dual 5-Input NOR Gate.
    - 744078: Single 8-input NOR Gate.

In the older RTL and ECL families, NOR gates were efficient and most commonly used.

Implementations

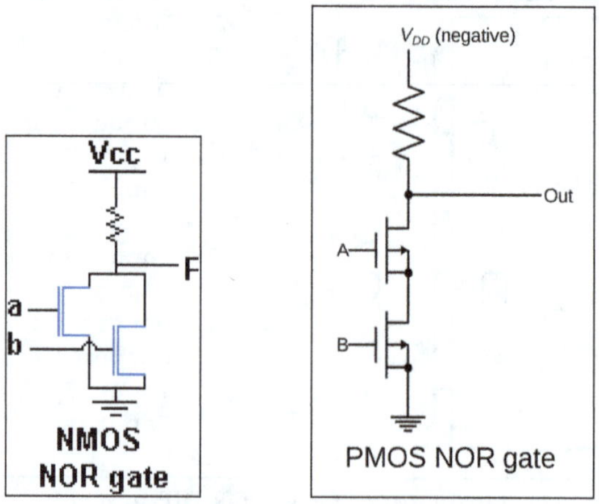

PMOS NOR gate with load resistor.

The diagrams above show the construction of a 2-input NOR gate using NMOS logic circuitry. If either of the inputs is high, the corresponding N-channel MOSFET is turned on and the output is pulled low; otherwise the output is pulled high through the pull-up resistor.

The diagram below shows a 2-input NOR gate using CMOS technology. The diodes and resistors on the inputs are to protect the CMOS components from damage due to electrostatic discharge (ESD) and play no part in the logical function of the circuit.

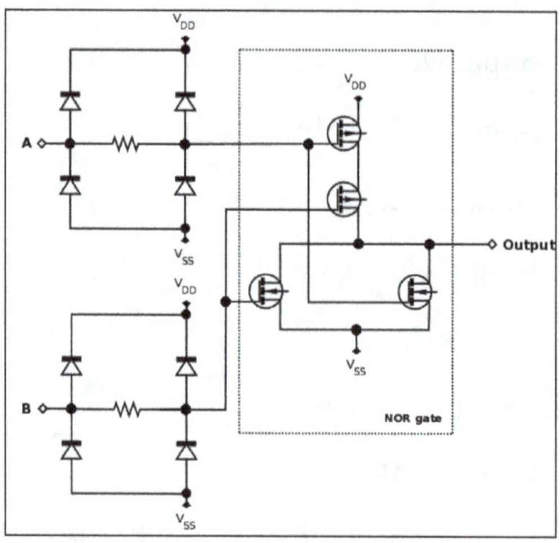

Unbuffered CMOS two input NOR gate.

Functional Completeness

The NOR gate has the property of functional completeness, which it shares with the NAND gate. That is, any other logic function (AND, OR, etc.) can be implemented using only NOR gates. An entire processor can be created using NOR gates alone.

As NAND gates are also functionally complete, if no specific NOR gates are available, one can be made from NAND gates.

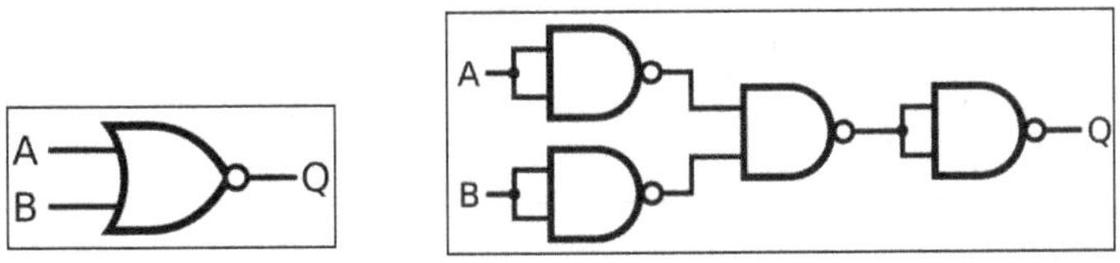

Desired gate. NAND Construction.

The above figure depicts a NAND construction of a NOR gate, which only implies the functional completeness of NOR under the correct assumption that NAND is functionally complete. However, a figure representing the decomposition of all the other logical functions into compositions of NORs would show this directly, without the assumption that NAND is functionally complete.

XOR Gate

XOR gate (sometimes EOR, or EXOR and pronounced as Exclusive OR) is a digital logic gate that gives a true (1 or HIGH) output when the number of true inputs is odd. An XOR gate implements an exclusive or; that is, a true output results if one, and only one, of the inputs to the gate is true. If both inputs are false (0/LOW) or both are true, a false output results. XOR represents the inequality function, i.e., the output is true if the inputs are not alike otherwise the output is false. A way to remember XOR is "must have one or the other but not both".

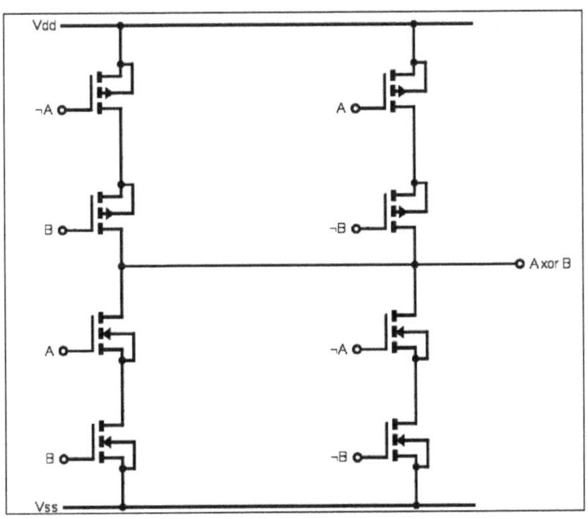

CMOS XOR gate.

INPUT		OUTPUT
A	B	A XOR B
0	0	0
0	1	1

1	0	1
1	1	0

XOR can also be viewed as addition modulo 2. As a result, XOR gates are used to implement binary addition in computers. A half adder consists of an XOR gate and an AND gate. Other uses include subtractors, comparators, and controlled inverters.

The algebraic expressions $A \cdot \overline{B} + \overline{A} \cdot B$ and $(A+B) \cdot (\overline{A} + \overline{B})$ both represent the XOR gate with inputs *A* and *B*. The behavior of XOR is summarized in the truth table shown on the right.

Symbols

There are two symbols for XOR gates: the traditional symbol and the IEEE symbol.

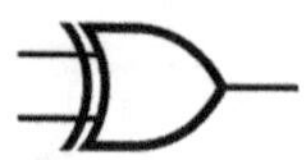

Traditional XOR Symbol.

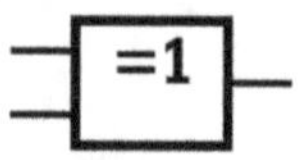

IEEE XOR Symbol.

The logic symbols ⊕, *Jpq*, and ⊻ can be used to denote XOR in algebraic expressions. C-like languages use the caret symbol ^ to denote bitwise XOR. (The caret does not denote logical conjunction (AND) in these languages, despite the similarity of symbol).

Pass-gate-logic Wiring

An XOR gate can be constructed using MOSFETs. Here is a diagram of a pass transistor logic implementation of an XOR gate.

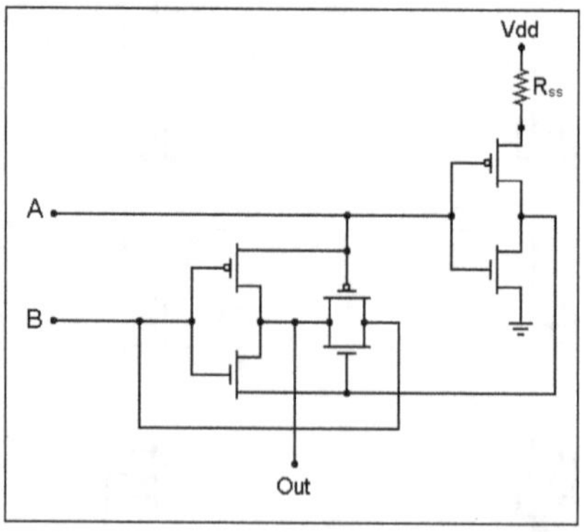

Transmission Gate Logic wiring of an XOR gate.

The "Rss" resistor prevents shunting current directly from "A" and "B" to the output. Without it, if the circuit that provides inputs A and B does not have the proper driving capability, the output might not swing rail to rail or be severely slew-rate limited. The "Rss" resistor also limits the current from Vdd to ground which protects the transistors and saves energy when the transistors are transitioning between states.

Analytical Representation

$f(a,b) = a + b - 2 * a * b$ is an analytical representation of XOR gate:

- $f(0,0) = 0 + 0 - 2 * 0 * 0 = 0$
- $f(0,1) = 0 + 1 - 2 * 0 * 1 = 1$
- $f(1,0) = 1 + 0 - 2 * 1 * 0 = 1$
- $f(1,1) = 1 + 1 - 2 * 1 * 1 = 0$

$f(a,b) = |a - b|$ is an alternative analytical representation.

Alternatives

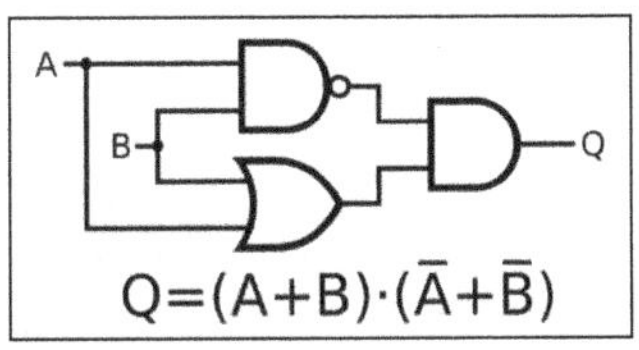

XOR gate circuit using three mixed gates.

If a specific type of gate is not available, a circuit that implements the same function can be constructed from other available gates. A circuit implementing an XOR function can be trivially constructed from an XNOR gate followed by a NOT gate. If we consider the expression $(A\cdot\overline{B})+(\overline{A}\cdot B)$, we can construct an XOR gate circuit directly using AND, OR and NOT gates. However, this approach requires five gates of three different kinds.

As alternative, if different gates are available we can apply Boolean algebra to transform $(A\cdot\overline{B})+(\overline{A}\cdot B) \equiv (A+B)\cdot(\overline{A}+\overline{B})$, and apply de Morgan's Law to the last term to get $(A+B)\cdot(\overline{A\cdot B})$ which can be implemented using only three gates as shown on the right.

An XOR gate circuit can be made from four NAND gates. In fact, both NAND and NOR gates are so-called "universal gates" and any logical function can be constructed from either NAND logic or NOR logic alone. If the four NAND gates are replaced by NOR gates, this results in an XNOR gate, which can be converted to an XOR gate by inverting the output or one of the inputs (e.g. with a fifth NOR gate).

Desired gate.

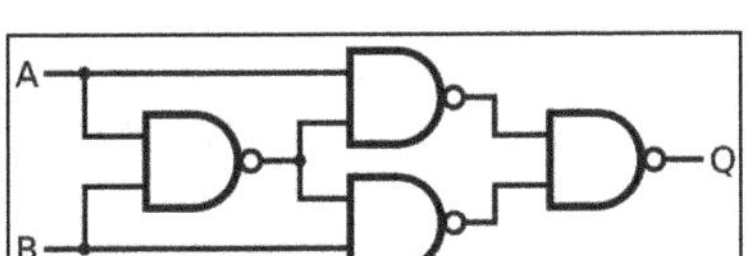

NAND construction.

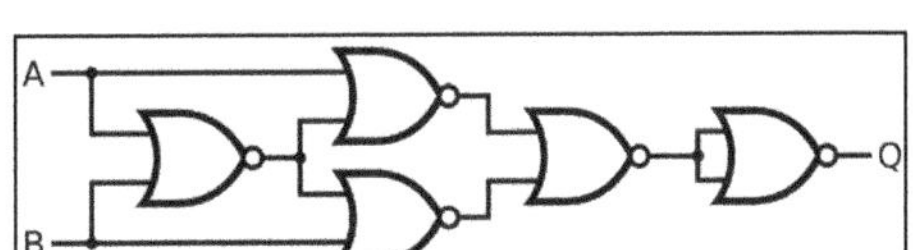

NOR construction.

An alternative arrangement is of five NOR gates in a topology that emphasizes the construction of the function from $(A+B)\cdot(\overline{A}+\overline{B})$, noting from de Morgan's Law that a NOR gate is an inverted-input AND gate. Another alternative arrangement is of five NAND gates in a topology that emphasizes the construction of the function from $(A \cdot \overline{B}) + (\overline{A} \cdot B)$, noting from de Morgan's Law that a NAND gate is an inverted-input OR gate.

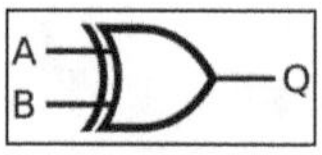

Desired gate.

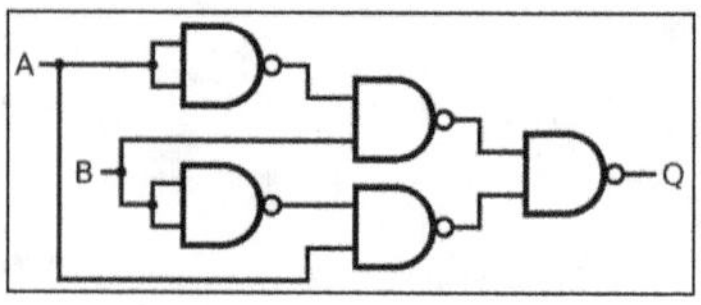

NAND construction.

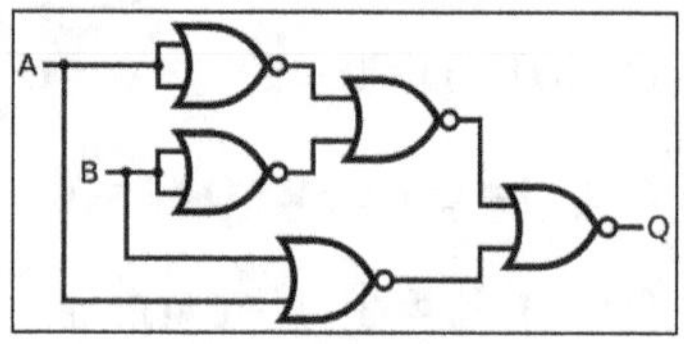

NOR construction.

For the NAND constructions, the upper arrangement requires fewer gates. For the NOR constructions, the lower arrangement offers the advantage of a shorter propagation delay (the time delay between an input changing and the output changing).

More than Two Inputs

Literal interpretation of the name "exclusive or", or observation of the IEC rectangular symbol, raises the question of correct behaviour with additional inputs. If a logic gate were to accept three or more inputs and produce a true output if exactly one of those inputs were true, then it would in effect be a one-hot detector (and indeed this is the case for only two inputs). However, it is rarely implemented this way in practice.

It is most common to regard subsequent inputs as being applied through a cascade of binary exclusive-or operations: the first two signals are fed into an XOR gate, then the output of that gate is fed into a second XOR gate together with the third signal, and so on for any remaining signals. The result is a circuit that outputs a 1 when the number of 1s at its inputs is odd, and a 0 when the number of incoming 1s is even. This makes it practically useful as a parity generator or a modulo-2 adder.

For example, the 74LVC1G386 microchip is advertised as a three-input logic gate, and implements a parity generator.

Applications

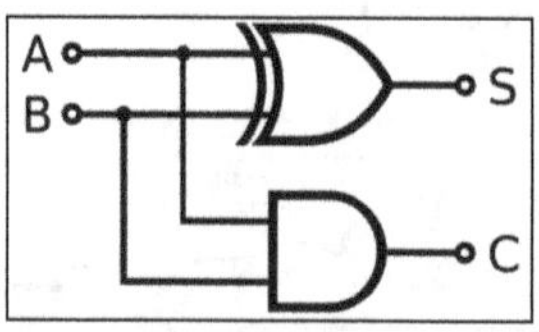

Example half adder circuit diagram.

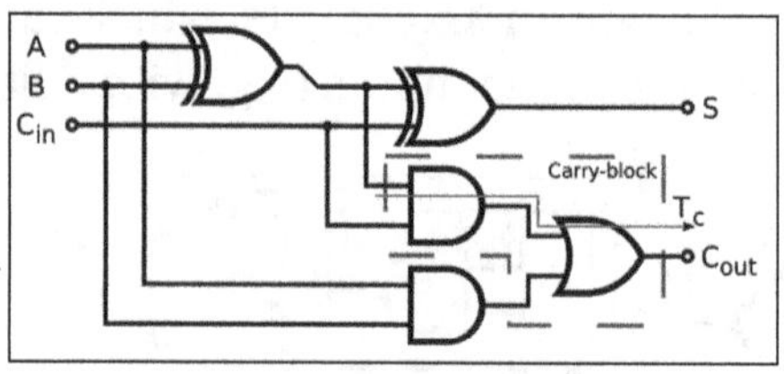

Example full adder circuit diagram.

Uses in Addition

The XOR logic gate can be used as a one-bit adder that adds any two bits together to output one bit. For example, if we add 1 plus 1 in binary, we expect a two-bit answer, 10 (i.e. 2 in decimal). Since the trailing sum bit in this output is achieved with XOR, the preceding carry bit is calculated with an AND gate. This is the main principle in Half Adders. A slightly larger Full Adder circuit may be chained together in order to add longer binary numbers.

Pseudo-random Number Generator

Pseudo-random number (PRN) generators, specifically Linear feedback shift registers, are defined in terms of the exclusive-or operation. Hence, a suitable setup of XOR gates can model a linear feedback shift register, in order to generate random numbers.

Correlation and Sequence Detection

XOR gates produce a 0 when both inputs match. When searching for a specific bit pattern or PRN sequence in a very long data sequence, a series of XOR gates can be used to compare a string of bits from the data sequence against the target sequence in parallel. The number of 0 outputs can then be counted to determine how well the data sequence matches the target sequence. Correlators are used in many communications devices such as CDMA receivers and decoders for error correction and channel codes. In a CDMA receiver, correlators are used to extract the polarity of a specific PRN sequence out of a combined collection of PRN sequences.

A correlator looking for 11010 in the data sequence 1110100101 would compare the incoming data bits against the target sequence at every possible offset while counting the number of matches (zeros):

```
1110100101 (data)

11010      (target)

00111      (XOR) 2 zero bits


1110100101

 11010

 00000     5 zero bits


1110100101

  11010

  01110    2 zero bits


1110100101

   11010

   10011   2 zero bits
```

```
1110100101
   11010
   01000   4 zero bits

1110100101
    11010
    11111 0 zero bits

Matches by offset:
  .
  :    :
: : : : :
-----------
0 1 2 3 4 5
```

In this example, the best match occurs when the target sequence is offset by 1 bit and all five bits match. When offset by 5 bits, the sequence exactly matches its inverse. By looking at the difference between the number of ones and zeros that come out of the bank of XOR gates, it is easy to see where the sequence occurs and whether or not it is inverted. Longer sequences are easier to detect than short sequences.

XNOR Gate

The XNOR gate (sometimes ENOR, EXNOR or NXOR and pronounced as Exclusive NOR) is a digital logic gate whose function is the logical complement of the exclusive OR (XOR) gate. The two-input version implements logical equality, behaving according to the truth table to the right, and hence the gate is sometimes called an "equivalence gate". A high output (1) results if both of the inputs to the gate are the same. If one but not both inputs are high (1), a low output (0) results.

Input		Output
A	B	A XNOR B
0	0	1
0	1	0
1	0	0
1	1	1

The algebraic notation used to represent the XNOR operation is $S = A \odot B$. The algebraic expressions $(A+\overline{B})\cdot(\overline{A}+B)$ and $A\cdot B+\overline{A}\cdot\overline{B}$ both represent the XNOR gate with inputs *A* and *B*.

Symbols

There are two symbols for XNOR gates: one with distinctive shape and one with rectangular shape and label. Both symbols for the XNOR gate are that of the XOR gate with an added inversion bubble.

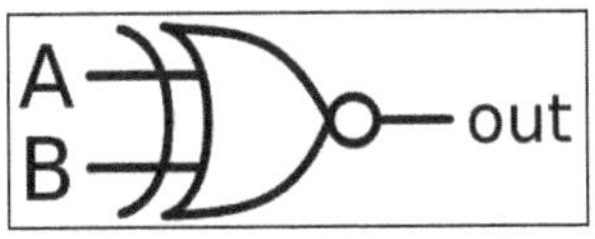

Distinctive symbol.

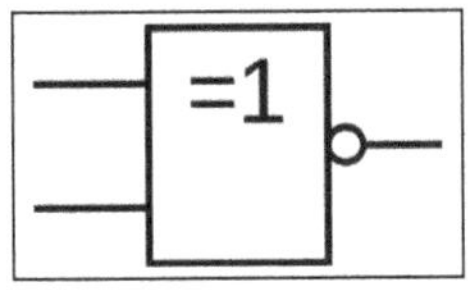

Rectangular symbol.

Hardware Description

XNOR gates are represented in most TTL and CMOS IC families. The standard 4000 series CMOS IC is the 4077, and the TTL IC is the 74266 (although an open-collector implementation). Both include four independent, two-input, XNOR gates. The (now obsolete) 74S135 implemented four two-input XOR/XNOR gates or two three-input XNOR gates.

Both the TTL 74LS implementation, the 74LS266, as well as the CMOS gates (CD4077, 74HC4077 and 74HC266 and so on) are available from most semiconductor manufacturers such as Texas Instruments or NXP, etc. They are usually available in both through-hole DIP and SOIC formats (SOIC-14, SOC-14 or TSSOP-14).

Datasheets are readily available in most datasheet databases and suppliers.

Pinout

Both the 4077 and 74x266 devices (SN74LS266, 74HC266, 74266, etc.) have the same pinout diagram, as follows:

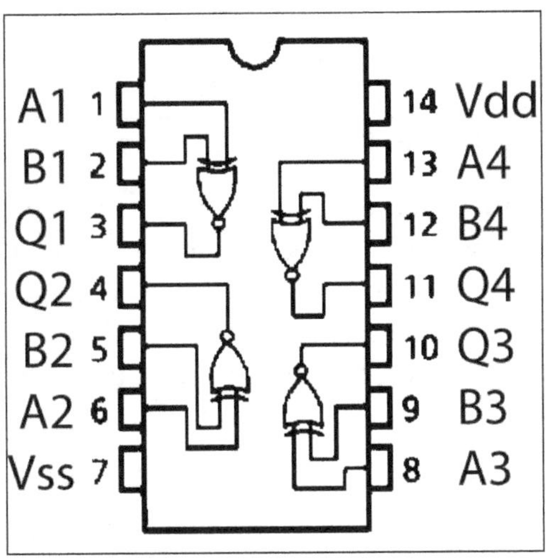

Pinout diagram of the 74HC266N, 74LS266 and CD4077 quad XNOR plastic dual in-line package 14-pin package (PDIP-14) ICs.

- Input A1.
- Input B1.
- Output Q1 (high if and only if A1 and B1 have the same logic level).
- Output Q2.
- Input B2.
- Input A2.
- V_{ss} (GND) common power and signal ground pin.
- Input A3.
- Input B3.
- Output Q3.
- Output Q4.
- Input B4.
- Input A4.
- V_{dd} for CMOS (V_{cc} for TTL) positive power supply.

Alternatives

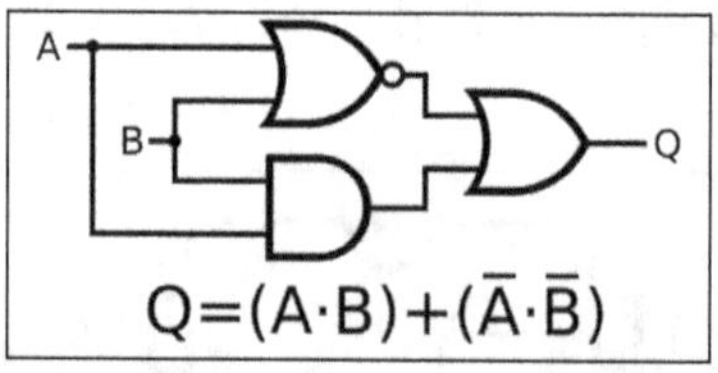

XNOR gate circuit using three mixed gates.

If a specific type of gate is not available, a circuit that implements the same function can be constructed from other available gates. A circuit implementing an XNOR function can be trivially constructed from an XOR gate followed by a NOT gate. If we consider the expression $(A+\overline{B})\cdot(\overline{A}+B)$, we can construct an XNOR gate circuit directly using AND, OR and NOT gates. However, this approach requires five gates of three different kinds.

As alternative, if different gates are available we can apply Boolean algebra to transform $(A+\overline{B})\cdot(\overline{A}+B)\equiv(A\cdot B)+(\overline{A}\cdot\overline{B})$ as stated above, and apply de Morgan's Law to the last term to get $(A \cdot B)+\overline{(A+B)}$ which can be implemented using only three gates as shown on the right.

An XNOR gate circuit can be made from four NOR gates. In fact, both NAND and NOR gates are so-called "universal gates" and any logical function can be constructed from either NAND logic or NOR logic alone. If the four NOR gates are replaced by NAND gates, this results in an XOR gate, which can be converted to an XNOR gate by inverting the output or one of the inputs (e.g. with a fifth NAND gate).

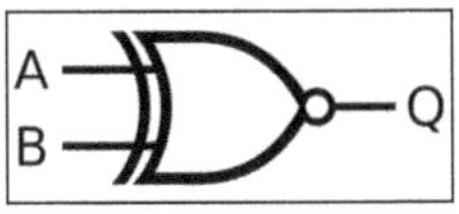

Desired gate.

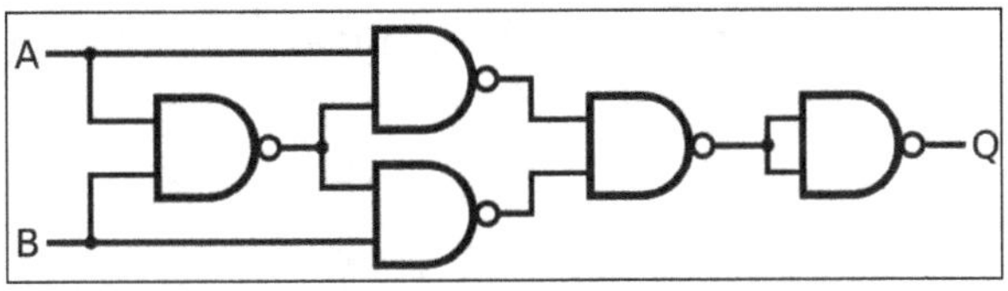

NAND construction.

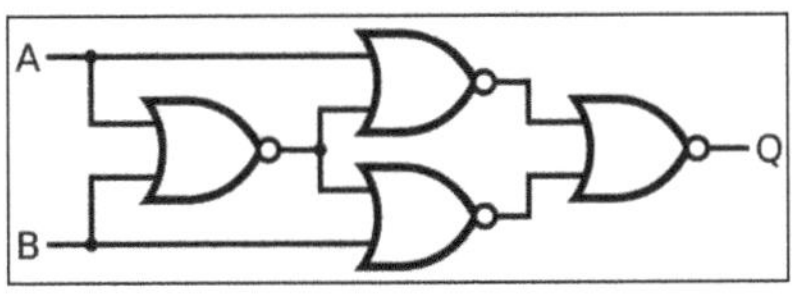

NOR construction.

An alternative arrangement is of five NAND gates in a topology that emphasizes the construction of the function from $(A \cdot B)+(\overline{A} \cdot \overline{B})$, noting from de Morgan's Law that a NAND gate is an inverted-input OR gate. Another alternative arrangement is of five NOR gates in a topology that emphasizes the construction of the function from $(A+\overline{B}) \cdot (\overline{A}+B)$, noting from de Morgan's Law that a NOR gate is an inverted-input AND gate.

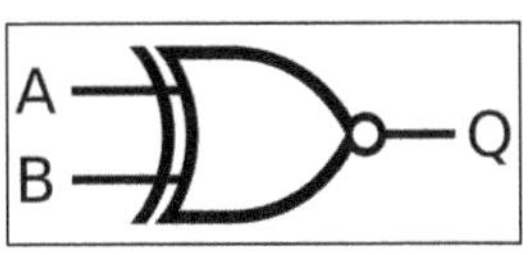

Desired gate.

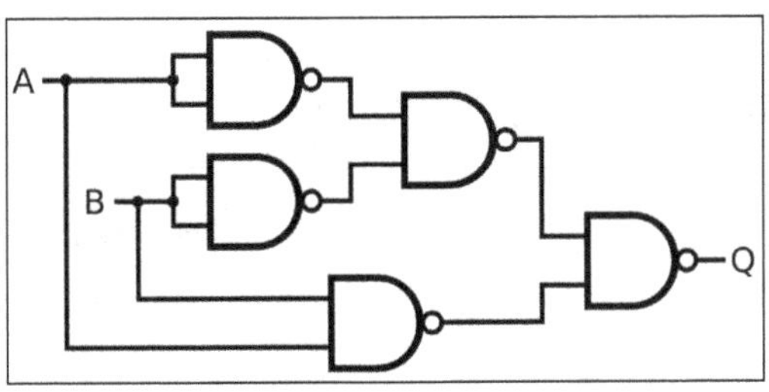

NAND construction.

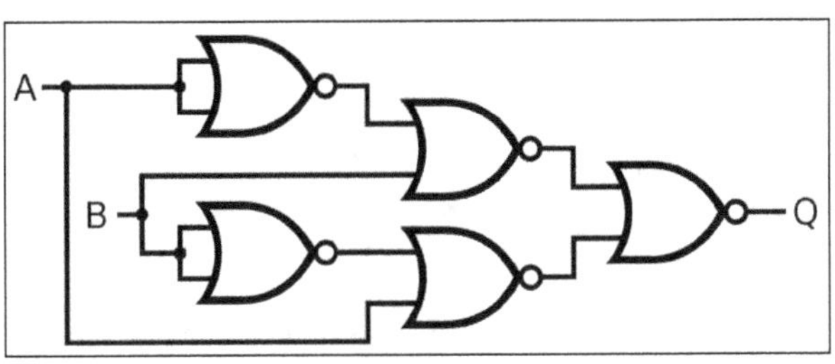

NOR construction.

For the NAND constructions, the lower arrangement offers the advantage of a shorter propagation delay (the time delay between an input changing and the output changing). For the NOR constructions, the upper arrangement requires fewer gates.

More than Two Inputs

Although other gates (OR, NOR, AND, NAND) are available from manufacturers with three or more inputs per gate, this is not strictly true with XOR and XNOR gates. However, extending the concept of the binary logical operation to three inputs, the SN74S135 with two shared "C" and four independent "A" and "B" inputs for its four outputs, was a device that followed the truth table:

Input			Output
A	B	C	Y
0	0	0	1
0	0	1	0
0	1	0	0
0	1	1	1
1	0	0	0
1	0	1	1
1	1	0	1
1	1	1	0

This is effectively Y = NOT ((A XOR B) XOR C). Another way to interpret this is that the output is true if an even number of inputs are true. It does not implement a logical "equivalence" function, unlike two-input XNOR gates.

Combinational Logic

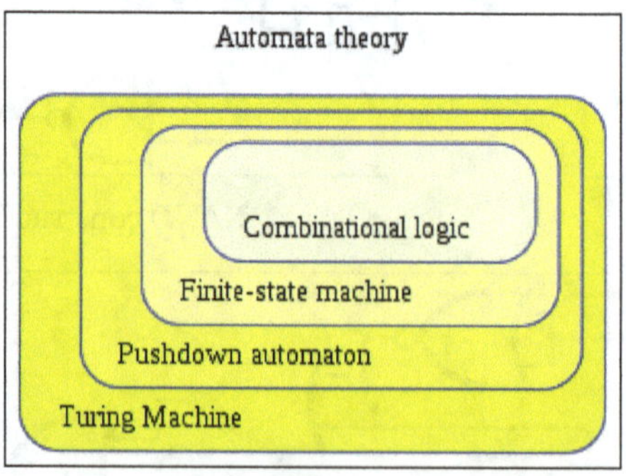

Classes of automata.

In digital circuit theory, combinational logic is a type of digital logic which is implemented by Boolean circuits, where the output is a pure function of the present input only. This is in contrast to sequential logic, in which the output depends not only on the present input but also on the history of the input. In other words, sequential logic has memory while combinational logic does not.

Combinational logic is used in computer circuits to perform Boolean algebra on input signals and on stored data. Practical computer circuits normally contain a mixture of combinational and sequential logic. For example, the part of an arithmetic logic unit, or ALU, that does mathematical calculations is constructed using combinational logic. Other circuits used in computers, such as half adders, full adders, half subtractors, full subtractors, multiplexers, demultiplexers, encoders and decoders are also made by using combinational logic.

Practical design of combinational logic systems may require consideration of the finite time required for practical logical elements to react to changes in their inputs. Where an output is the result of the combination of several different paths with differing numbers of switching elements, the output may momentarily change state before settling at the final state, as the changes propagate along different paths.

An alternate term is combinatorial logic.

Representation

Combinational logic is used to build circuits that produce specified outputs from certain inputs. The construction of combinational logic is generally done using one of two methods: a sum of products, or a product of sums. Consider the following truth table :

A	B	C	Result	Logical equivalent
F	F	F	F	$\neg A \cdot \neg B \cdot \neg C$
F	F	T	F	$\neg A \cdot \neg B \cdot C$
F	T	F	F	$\neg A \cdot B \cdot \neg C$
F	T	T	F	$\neg A \cdot B \cdot C$
T	F	F	T	$A \cdot \neg B \cdot \neg C$
T	F	T	F	$A \cdot \neg B \cdot C$
T	T	F	F	$A \cdot B \cdot \neg C$
T	T	T	T	$A \cdot B \cdot C$

Using sum of products, all logical statements which yield true results are summed, giving the result:

$$A \cdot \neg B \cdot \neg C + A \cdot B \cdot C$$

Using Boolean algebra, the result simplifies to the following equivalent of the truth table:

$$A \cdot (\neg B \cdot \neg C + B \cdot C)$$

Logic Formula Minimization

Minimization (simplification) of combinational logic formulas is done using the following rules based on the laws of Boolean algebra:

$$(A+B)\cdot(A+C)=A+(B\cdot C)$$
$$(A\cdot B)+(A\cdot C)=A\cdot(B+C)$$

$$A+(A\cdot B)=A$$
$$A\cdot(A+B)=A$$

$$A+(\neg A\cdot B)=A+B$$
$$A\cdot(\neg A+B)=A\cdot B$$

$$(A+B)\cdot(\neg A+B)=B$$
$$(A\cdot B)+(\neg A\cdot B)=B$$

$$(A\cdot B)+(\neg A\cdot C)+(B\cdot C)=(A\cdot B)+(\neg A\cdot C)$$
$$(A+B)\cdot(\neg A+C)\cdot(B+C)=(A+B)\cdot(\neg A+C)$$

With the use of minimization (sometimes called logic optimization), a simplified logical function or circuit may be arrived upon, and the logic combinational circuit becomes smaller, and easier to analyse, use, or build.

Sequential Logic

In digital circuit theory, sequential logic is a type of logic circuit whose output depends not only on the present value of its input signals but on the sequence of past inputs, the input history as well. This is in contrast to combinational logic, whose output is a function of only the present input. That is, sequential logic has state (memory) while combinational logic does not.

Sequential logic is used to construct finite state machines, a basic building block in all digital circuitry. Virtually all circuits in practical digital devices are a mixture of combinational and sequential logic.

A familiar example of a device with sequential logic is a television set with "channel up" and "channel down" buttons. Pressing the "up" button gives the television an input telling it to switch to the next channel above the one it is currently receiving. If the television is on channel 5, pressing "up" switches it to receive channel 6. However, if the television is on channel 8, pressing "up" switches it to channel "9". In order for the channel selection to operate correctly, the television must be aware of which channel it is currently receiving, which was determined by past channel selections. The television stores the current channel as part of its state. When a "channel up" or "channel down" input is given to it, the sequential logic of the channel selection circuitry calculates the new channel from the input and the current channel.

Digital sequential logic circuits are divided into synchronous and asynchronous types. In synchronous sequential circuits, the state of the device changes only at discrete times in response to a clock signal. In asynchronous circuits the state of the device can change at any time in response to changing inputs.

Synchronous Circuit

A synchronous circuit is a digital circuit in which the changes in the state of memory elements are synchronized by a clock signal. In a sequential digital logic circuit, data is stored in memory devices called flip-flops or latches. The output of a flip-flop is constant until a pulse is applied to its "clock" input, upon which the input of the flip-flop is latched into its output. In a synchronous logic circuit, an electronic oscillator called the *clock* generates a string of pulses, the "clock signal". This clock signal is applied to every storage element, so in an ideal synchronous circuit, every change in the logical levels of its storage components is simultaneous. Ideally, the input to each storage element has reached its final value before the next clock occurs, so the behaviour of the whole circuit can be predicted exactly. Practically, some delay is required for each logical operation, resulting in a maximum speed at which each synchronous system can run.

To make these circuits work correctly, a great deal of care is needed in the design of the clock distribution networks. Static timing analysis is often used to determine the maximum safe operating speed.

Nearly all digital circuits, and in particular nearly all CPUs, are fully synchronous circuits with a global clock. Exceptions are often compared to fully synchronous circuits. Exceptions include self-synchronous circuits, globally asynchronous locally synchronous circuits, and fully asynchronous circuits.

Asynchronous Circuit

An asynchronous circuit, or self-timed circuit, is a sequential digital logic circuit which is not governed by a clock circuit or global clock signal. Instead it often uses signals that indicate completion of instructions and operations, specified by simple data transfer protocols. This type of circuit is contrasted with synchronous circuits, in which changes to the signal values in the circuit are triggered by repetitive pulses called a clock signal. Most digital devices today use synchronous circuits. However asynchronous circuits have the potential to be faster, and may also have advantages in lower power consumption, lower electromagnetic interference, and better modularity in large systems. Asynchronous circuits are an active area of research in digital logic design.

Synchronous vs. Asynchronous Logic

Digital logic circuits can be divided into combinational logic, in which the output signals depend only on the current input signals, and sequential logic, in which the output depends both on current input and on past inputs. In other words, sequential logic is combinational logic with memory. Virtually all practical digital devices require sequential logic. Sequential logic can be divided into two types, synchronous logic and asynchronous logic:

- In synchronous logic circuits, an electronic oscillator generates a repetitive series of equally spaced pulses called the clock signal. The clock signal is applied to all the memory elements in the circuit, called flip-flops. The output of the flip-flops only changes when triggered by the edge of the clock pulse, so changes to the logic signals throughout the circuit all begin at the same time, at regular intervals synchronized by the clock. The output of all memory elements in a circuit is called the state of the circuit. The state of a synchronous circuit changes only on the clock pulse. The changes in signal require a certain amount of time to propagate through the combinational logic gates of the circuit. This is called propagation delay. The period of the clock signal is made long enough so the output of all the logic gates have time to settle to stable values before the next clock pulse. As long as this condition is met, synchronous circuits will operate stably, so they are easy to design.

  However a disadvantage of synchronous circuits is that they can be slow. The maximum possible clock rate is determined by the logic path with the longest propagation delay, called the critical path. So logic paths that complete their operations quickly are idle most of the time. Another problem is that the widely distributed clock signal takes a lot of power, and must run whether the circuit is receiving inputs or not.

- In asynchronous circuits, there is no clock signal, and the state of the circuit changes as soon as the inputs change. Since asynchronous circuits don't have to wait for a clock pulse to begin processing inputs, they can be faster than synchronous circuits, and their speed is theoretically limited only by the propagation delays of the logic gates. However, asynchronous circuits are more difficult to design and subject to problems not found in synchronous circuits. This is because the resulting state of an asynchronous circuit can be sensitive to the relative arrival times of inputs at gates. If transitions on two inputs arrive at almost the same time, the circuit can go into the wrong state depending on slight differences in the propagation delays of the gates. This is called a race condition. In synchronous circuits this problem is less severe because race conditions can only occur due to inputs from outside the synchronous system, called asynchronous inputs. Although some fully asynchronous digital systems have been built, today asynchronous circuits are typically used in a few critical parts of otherwise synchronous systems where speed is at a premium, such as signal processing circuits.

Benefits

A variety of advantages have been demonstrated by asynchronous circuits, including both quasi-delay-insensitive (QDI) circuits (generally agreed to be the most "pure" form of asynchronous logic that retains computational universality) and less pure forms of asynchronous circuitry which use timing constraints for higher performance and lower area and power:

- Robust handling of metastability of arbiters.
- Higher performance function units, which provide average-case (i.e. data-dependent) completion rather than worst-case completion. Examples include speculative completion which has been applied to design parallel prefix adders faster than synchronous ones, and a high-performance double-precision floating point adder' which outperforms leading synchronous designs.
- Early completion of a circuit when it is known that the inputs which have not yet arrived are irrelevant.
- Lower power consumption because no transistor ever transitions unless it is performing useful computation. Epson has reported 70% lower power consumption compared to synchronous design. Also, clock drivers can be removed which can significantly reduce power consumption. However, when using certain encodings, asynchronous circuits may require more area, which can result in increased power consumption if the underlying process has poor leakage properties (for example, deep submicrometer processes used prior to the introduction of High-k dielectrics).
- "Elastic" pipelines, which achieve high performance while gracefully handling variable input and output rates and mismatched pipeline stage delays.
- Freedom from the ever-worsening difficulties of distributing a high-fan-out, timing-sensitive clock signal.
- Better modularity and composability.
- Far fewer assumptions about the manufacturing process are required (most assumptions are timing assumptions).
- Circuit speed adapts to changing temperature and voltage conditions rather than being locked at the speed mandated by worst-case assumptions.
- Immunity to transistor-to-transistor variability in the manufacturing process, which is one of the most serious problems facing the semiconductor industry as dies shrink.
- Less severe electromagnetic interference (EMI). Synchronous circuits create a great deal of EMI in the frequency band at (or very near) their clock frequency

and its harmonics; asynchronous circuits generate EMI patterns which are much more evenly spread across the spectrum.

- In asynchronous circuits, local signaling eliminates the need for global synchronization which exploits some potential advantages in comparison with synchronous ones. They have shown potential specifications in low power consumption, design reuse, improved noise immunity and electromagnetic compatibility. Asynchronous circuits are more tolerant to process variations and external voltage fluctuations.
- Less stress on the power distribution network. Synchronous circuits tend to draw a large amount of current right at the clock edge and shortly thereafter. The number of nodes switching (and thence, amount of current drawn) drops off rapidly after the clock edge, reaching zero just before the next clock edge. In an asynchronous circuit, the switching times of the nodes are not correlated in this manner, so the current draw tends to be more uniform and less bursty.

Disadvantages

- Area overhead caused by an increase in the number of circuit elements (transistors). In some cases an asynchronous design may require up to double the resources of a synchronous design, due to addition of completion detection and design-for-test circuits.
- Fewer people are trained in this style compared to synchronous design.
- Synchronous designs are inherently easier to test and debug than asynchronous designs. However, this position is disputed by Fant, who claims that the apparent simplicity of synchronous logic is an artifact of the mathematical models used by the common design approaches.
- Clock gating in more conventional synchronous designs is an approximation of the asynchronous ideal, and in some cases, its simplicity may outweigh the advantages of a fully asynchronous design.
- Performance (speed) of asynchronous circuits may be reduced in architectures that require input-completeness (more complex data path).
- Lack of dedicated, asynchronous design-focused commercial EDA tools.

Communication

There are several ways to create asynchronous communication channels that can be classified by their protocol and data encoding.

Protocols

There are two widely used protocol families which differ in the way communications are encoded:

- Two-phase handshake (a.k.a. two-phase protocol, Non-Return-to-Zero (NRZ) encoding, or transition signalling): Communications are represented by any wire transition; transitions from 0 to 1 and from 1 to 0 both count as communications.
- Four-phase handshake (a.k.a. four-phase protocol, or Return-to-Zero (RZ) encoding): Communications are represented by a wire transition followed by a reset; a transition sequence from 0 to 1 and back to 0 counts as single communication.

Illustration of two and four-phase handshakes: Top: A sender and a receiver are communicating with simple request and acknowledge signals. The sender drives the request line, and the receiver drives the acknowledge line. Middle: Timing diagram of two, two-phase communications. Bottom: Timing diagram of one, four-phase communication.

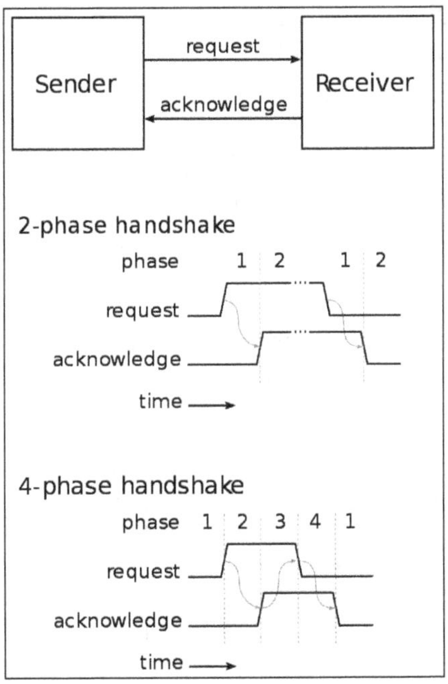

Despite involving more transitions per communication, circuits implementing four-phase protocols are usually faster and simpler than two-phase protocols because the signal lines return to their original state by the end of each communication. In two-phase protocols, the circuit implementations would have to store the state of the signal line internally.

These basic distinctions do not account for the wide variety of protocols. These protocols may encode only requests and acknowledgements or also encode the data, which leads to the popular multi-wire data encoding. Many other, less common protocols have been proposed including using a single wire for request and acknowledgment, using several significant voltages, using only pulses or balancing timings in order to remove the latches.

Data Encoding

There are two widely used data encodings in asynchronous circuits: bundled-data encoding and multi-rail encoding.

Another common way to encode the data is to use multiple wires to encode a single digit: the value is determined by the wire on which the event occurs. This avoids some of the delay assumptions necessary with bundled-data encoding, since the request and the data are not separated anymore.

Bundled-data Encoding

Bundled-data encoding uses one wire per bit of data with a request and an acknowledge signal; this is the same encoding used in synchronous circuits without the restriction that transitions occur on with a clock edge. The request and the acknowledge are sent on separate wires with one of the above protocols. These circuits usually assume a bounded delay model with the completion signals delayed long enough for the calculations to take place.

In operation, the sender signals the availability and validity of data with a request. The receiver then indicates completion with an acknowledgement, indicating that it is able to process new requests. That is, the request is bundled with the data, hence the name "bundled-data".

Bundled-data circuits are often referred to as micropipelines, whether they use a two-phase or four-phase protocol, even if the term was initially introduced for two-phase bundled-data.

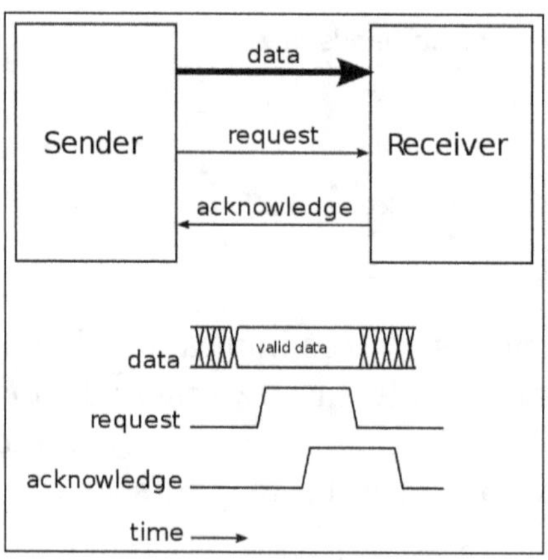

A 4-phase, bundled-data communication: Top: A sender and receiver are connected by data lines, a request line, and an acknowledge line. Bottom: Timing diagram of a bundled data communication. When the request line is low, the data is to be considered invalid and liable to change at any time.

Multi-rail Encoding

Multi-rail encoding uses multiple wires without a one-to-one relationship between bits and wires and a separate acknowledge signal. Data availability is indicated by the transitions themselves on one or more of the data wires (depending on the type of multi-rail encoding) instead of with a request signal as in the bundled-data encoding. This provides the advantage that the data communication is delay-insensitive. Two common multi-rail encodings are one-hot and dual rail. The one-hot (a.k.a. 1-of-n) encoding represents a number in base n with a communication on one of the n wires. The dual-rail encoding uses pairs of wires to represent each bit of the data, hence the name "dual-rail"; one wire in the pair represents the bit value of 0 and the other represents the bit value of 1. For example, a dual-rail encoded two bit number will be represented with two pairs of wires for four wires in total. During a data communication, communications occur on one of each pair of wires to indicate the data's bits. In the general case, an m × n encoding represent data as m words of base n.

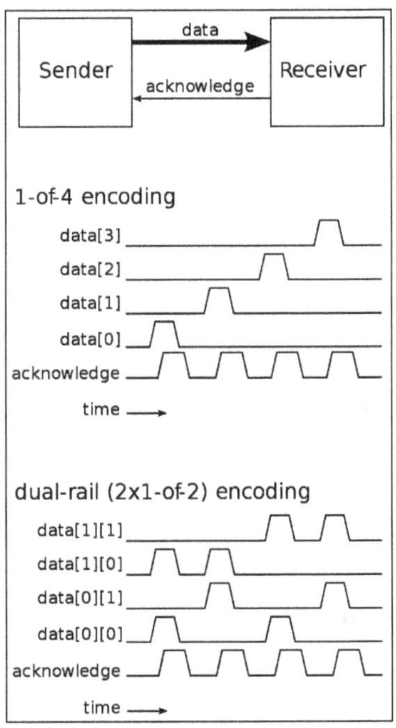

Diagram of dual rail and 1-of-4 communications: Top: A sender and receiver are connected by data lines and an acknowledge line. Middle: Timing diagram of the sender communicating the values 0, 1, 2, and then 3 to the receiver with the 1-of-4 encoding. Bottom: Timing diagram of the sender communicating the same values to the receiver with the dual-rail encoding. For this particular data size, the dual rail encoding is the same as a 2 x 1-of-2 encoding.

Dual-rail encoding with a four-phase protocol is the most common and is also called three-state encoding, since it has two valid states (10 and 01, after a transition) and a reset state (00). Another common encoding, which leads to a simpler implementation than one-hot, two-phase dual-rail is four-state encoding, or level-encoded dual-rail, and uses a data bit and a parity bit to achieve a two-phase protocol.

Asynchronous CPU

Asynchronous CPUs are one of several ideas for radically changing CPU design.

Unlike a conventional processor, a clockless processor (asynchronous CPU) has no central clock to coordinate the progress of data through the pipeline. Instead, stages of the CPU are coordinated using logic devices called "pipeline controls" or "FIFO sequencers." Basically, the pipeline controller clocks the next stage of logic when the existing stage is complete. In this way, a central clock is unnecessary. It may actually be even easier to implement high performance devices in asynchronous, as opposed to clocked, logic:

- Components can run at different speeds on an asynchronous CPU; all major components of a clocked CPU must remain synchronized with the central clock.
- A traditional CPU cannot "go faster" than the expected worst-case performance of the slowest stage/instruction/component. When an asynchronous CPU completes an operation more quickly than anticipated, the next stage can immediately begin processing the results, rather than waiting for synchronization with a central clock. An operation might finish faster than normal because of attributes of the data being processed (e.g., multiplication can be very fast when multiplying by 0 or 1, even when running code produced by a naive compiler), or because of the presence of a higher voltage or bus speed setting, or a lower ambient temperature, than 'normal' or expected.

Asynchronous logic proponents believe these capabilities would have these benefits:

- Lower power dissipation for a given performance level.
- Highest possible execution speeds.

The biggest disadvantage of the clockless CPU is that most CPU design tools assume a clocked CPU (i.e., a synchronous circuit). Many tools "enforce synchronous design practices". Making a clockless CPU (designing an asynchronous circuit) involves modifying the design tools to handle clockless logic and doing extra testing to ensure the design avoids metastable problems. The group that designed the AMULET, for example, developed a tool called LARD to cope with the complex design of AMULET3.

The ILLIAC II was the first completely asynchronous, speed independent processor design ever built; it was the most powerful computer at the time.

DEC PDP-16 Register Transfer Modules allowed the experimenter to construct asynchronous, 16-bit processing elements. Delays for each module were fixed and based on the module's worst-case timing.

The Caltech Asynchronous Microprocessor was the first asynchronous microprocessor. Caltech designed and manufactured the world's first fully Quasi Delay Insensitive processor. During demonstrations, the researchers loaded a simple program which ran in a tight loop, pulsing one of the output lines after each instruction. This output line was connected to an oscilloscope. When a cup of hot coffee was placed on the chip, the pulse rate (the effective "clock rate") naturally slowed down to adapt to the worsening performance of the heated transistors. When liquid nitrogen was poured on the chip, the instruction rate shot up with no additional intervention. Additionally, at lower temperatures, the voltage supplied to the chip could be safely increased, which also improved the instruction rate – again, with no additional configuration.

In 2004, Epson manufactured the world's first bendable microprocessor called ACT11, an 8-bit asynchronous chip. Synchronous flexible processors are slower, since bending the material on which a chip is fabricated causes wild and unpredictable variations in the delays of various transistors, for which worst-case scenarios must be assumed everywhere and everything must be clocked at worst-case speed. The processor is intended for use in smart cards, whose chips are currently limited in size to those small enough that they can remain perfectly rigid. In 2014, IBM announced a SyNAPSE-developed chip that runs in an asynchronous manner, with one of the highest transistor counts of any chip ever produced. IBM's chip consumes orders of magnitude less power than traditional computing systems on pattern recognition benchmarks.

Flip-flops

In electronics, a flip-flop or latch is a circuit that has two stable states and can be used to store state information – a bistable multivibrator. The circuit can be made to change state by signals applied to one or more control inputs and will have one or two outputs. It is the basic storage element in sequential logic. Flip-flops and latches are fundamental building blocks of digital electronics systems used in computers, communications, and many other types of systems.

We can implement flip-flops in two methods:

- In first method, cascade two latches in such a way that the first latch is enabled for every positive clock pulse and second latch is enabled for every negative clock pulse. So that the combination of these two latches become a flip-flop.
- In second method, we can directly implement the flip-flop, which is edge sensitive. In this chapter, let us discuss the following flip-flops using second method:
    - SR Flip-flop.

- D Flip-flop.
- JK Flip-flop.
- T Flip-flop.

SR Flip-flop

SR flip-flop operates with only positive clock transitions or negative clock transitions. Whereas, SR latch operates with enable signal. The circuit diagram of SR flip-flop is shown in the following figure:

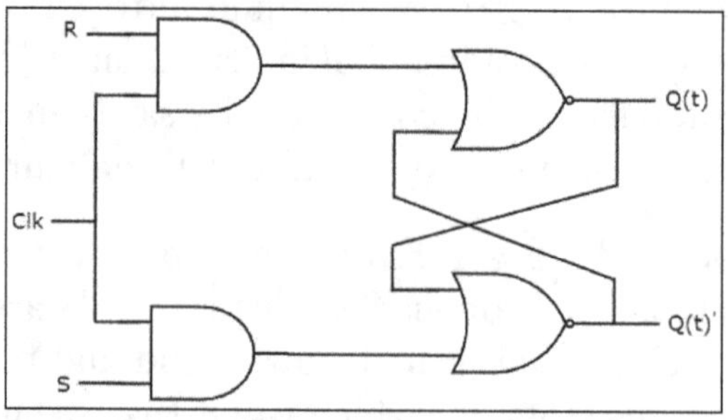

This circuit has two inputs S & R and two outputs Q t & Q t'. The operation of SR flipflop is similar to SR Latch. But, this flip-flop affects the outputs only when positive transition of the clock signal is applied instead of active enable.

The following table shows the state table of SR flip-flop:

S	R	Q t+1
0	0	Q t
0	1	0
1	0	1
1	1	-

Here, Q t & Q t+1 are present state & next state respectively. So, SR flip-flop can be used for one of these three functions such as Hold, Reset & Set based on the input conditions, when positive transition of clock signal is applied. The following table shows the characteristic table of SR flip-flop:

Present Inputs		Present State	Next State
S	R	Q t	Q t+1
0	0	0	0
0	0	1	1
0	1	0	0
0	1	1	0
1	0	0	1
1	0	1	1

1	1	0	x
1	1	1	x

By using three variables K-Map, we can get the simplified expression for next state, Qt+1.

The three variables K-Map for next state, Q t+1 is shown in the following figure:

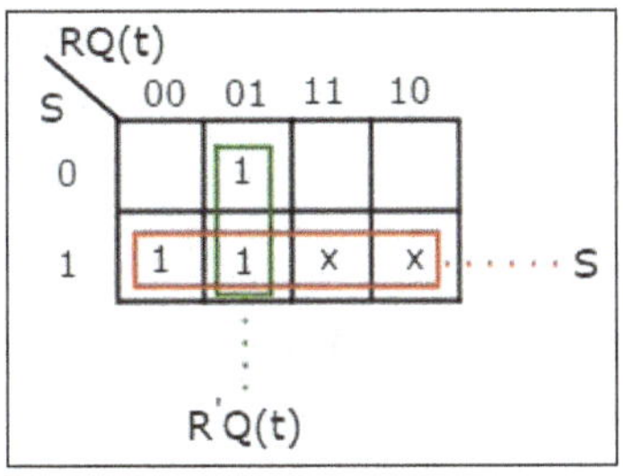

The maximum possible groupings of adjacent ones are already shown in the figure. Therefore, the simplified expression for next state Q t+1 is,

$$Q(t+1) = S + R'Q(t)$$

D Flip-flop

D flip-flop operates with only positive clock transitions or negative clock transitions. Whereas, D latch operates with enable signal. That means, the output of D flip-flop is insensitive to the changes in the input, D except for active transition of the clock signal. The circuit diagram of D flip-flop is shown in the following figure:

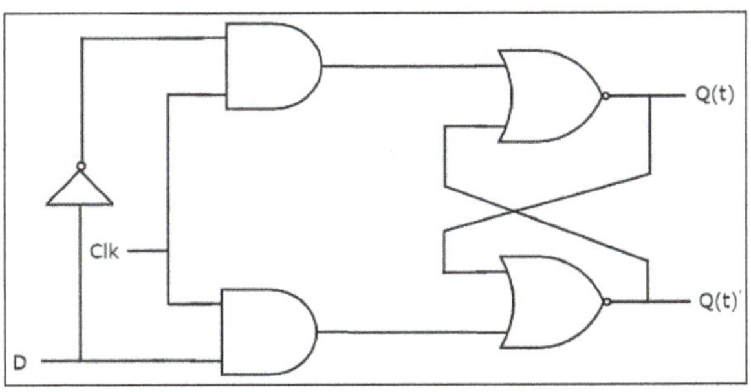

This circuit has single input D and two outputs Q t & Q t'. The operation of D flip-flop is similar to D Latch. But, this flip-flop affects the outputs only when positive transition of the clock signal is applied instead of active enable.

The following table shows the state table of D flip-flop:

D	Q t + 1t + 1
0	0
1	1

Therefore, D flip-flop always Hold the information, which is available on data input, D of earlier positive transition of clock signal. From the above state table, we can directly write the next state equation as,

$$Q\,t+1 = D$$

Next state of D flip-flop is always equal to data input, D for every positive transition of the clock signal. Hence, D flip-flops can be used in registers, shift registers and some of the counters.

JK Flip-flop

JK flip-flop is the modified version of SR flip-flop. It operates with only positive clock transitions or negative clock transitions. The circuit diagram of JK flip-flop is shown in the following figure:

This circuit has two inputs J & K and two outputs Q t & Q t'. The operation of JK flip-flop is similar to SR flip-flop. Here, we considered the inputs of SR flip-flop as S = J Q t' and R = KQ t in order to utilize the modified SR flip-flop for 4 combinations of inputs.

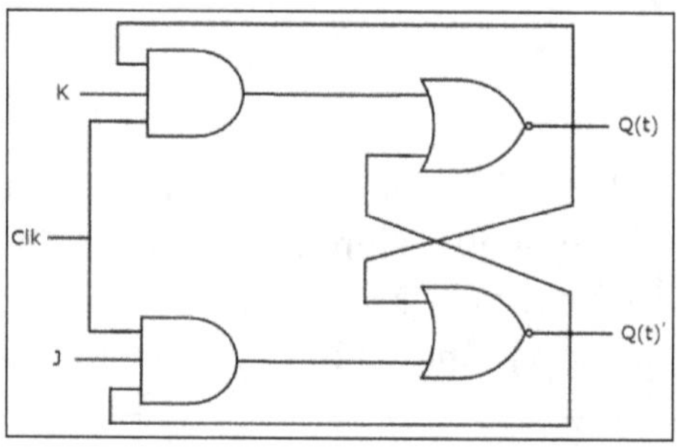

The following table shows the state table of JK flip-flop:

J	K	Q t+1
0	0	Q t
0	1	0
1	0	1
1	1	Q t'

Here, Q t & Q t+1 are present state & next state respectively. So, JK flip-flop can be used for one of these four functions such as Hold, Reset, Set & Complement of present state based on the input conditions, when positive transition of clock signal is applied. The following table shows the characteristic table of JK flip-flop:

Present Inputs		Present State	Next State
J	K	Q t	Q t+1
0	0	0	0

0	0	1	1
0	1	0	0
0	1	1	0
1	0	0	1
1	0	1	1
1	1	0	1
1	1	1	0

By using three variables K-Map, we can get the simplified expression for next state, $Q\ t+1$.

Three variables K-Map for next state, Q t+1 is shown in the following figure:

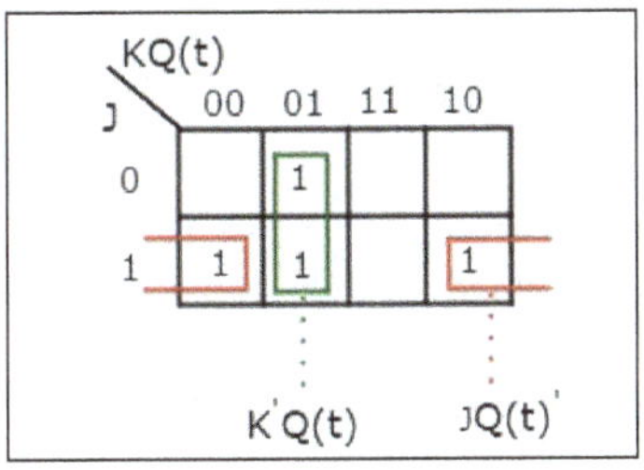

The maximum possible groupings of adjacent ones are already shown in the figure. Therefore, the simplified expression for next state Q t+1 is,

$$Q(t+1) = JQ(t)' + K'Q(t)$$

T Flip-flop

T flip-flop is the simplified version of JK flip-flop. It is obtained by connecting the same input 'T' to both inputs of JK flip-flop. It operates with only positive clock transitions or negative clock transitions. The circuit diagram of T flip-flop is shown in the following figure:

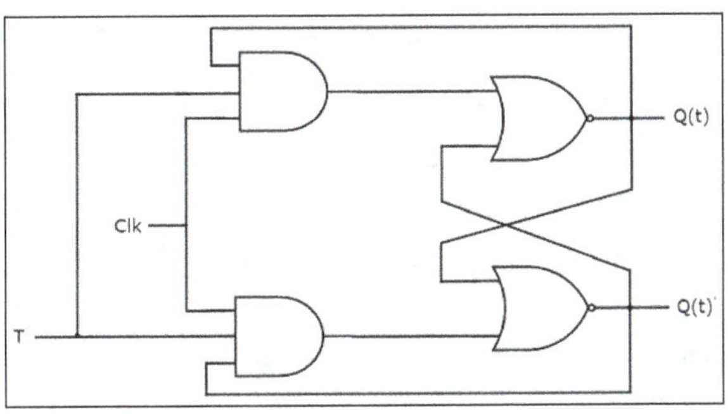

This circuit has single input T and two outputs Q t & Q t'. The operation of T flip-flop is same as that of JK flip-flop. Here, we considered the inputs of JK flip-flop as J = T and K = T in order to utilize the modified JK flip-flop for 2 combinations of inputs. So, we eliminated the other two combinations of J & K, for which those two values are complement to each other in T flip-flop.

The following table shows the state table of T flip-flop:

D	Q t+1
0	Q t
1	Q t'

Here, Q t & Q t+1 are present state & next state respectively. So, T flip-flop can be used for one of these two functions such as Hold, & Complement of present state based on the input conditions, when positive transition of clock signal is applied. The following table shows the characteristic table of T flip-flop:

Inputs	Present State	Next State
T	Q t	Q t+1
0	0	0
0	1	1
1	0	1
1	1	0

From the above characteristic table, we can directly write the next state equation as,

$$Q(t+1)=T'Q(t)+TQ(t)'$$
$$\Rightarrow Q(t+1)=T\oplus Q(t)$$

The output of T flip-flop always toggles for every positive transition of the clock signal, when input T remains at logic High 1. Hence, T flip-flop can be used in counters.

References

- Digital-electronics: electronicscoach.com, Retrieved 25 April, 2020
- Number-system-in-digital-electronics: circuitglobe.com, Retrieved 18 June, 2020
- M. Morris, Mano; R. Kime, Charles (2004). Logic and computer design fundamentals (3 ed.). Prentice Hall. p. 73. ISBN 0133760634
- Cavanagh, Joseph (2006). Sequential Logic: Analysis and Synthesis. CRC Press. pp. ix. ISBN 0849375649
- Logic-gate-AND-OR-XOR-NOT-NAND-NOR-and-XNOR, definition: whatis.techtarget.com, Retrieved 09 January, 2020
- Digital-circuits-flip-flops, digital-circuits: tutorialspoint.com, Retrieved 15 March, 2020
- Brumbach, Michael E. Industrial electricity (8th ed.). Clifton Park, N.Y.: Delmar. p. 546. ISBN 9781435483743

3
Analog Electronics

Analog electronics are the electronic systems that use continuous variable signals. Analogue filter, amplifier, bipolar junction transistor, field-effect transistor, etc. fall under its domain. The topics elaborated in this chapter will help in gaining a better perspective of the related aspects of analog electronics.

Analog electronics is a branch of electronics that deals with a continuously variable signal. It's widely used in radio and audio equipment along with other applications where signals are derived from analog sensors before being converted into digital signals for subsequent storage and processing. Although digital circuits are considered as a dominant part of today's technological world, some of the most fundamental components in a digital system are actually analog in nature.

Analog Signals

Signals are passed among devices in order to obtain or send information in the form of audio, video or encoded data. The transmission takes place through wires or via air through radio frequency waves. For instance, audio signals are transferred from the computer's audio card to the speakers, while data signals between a tablet and a Wi-Fi router pass through the air.

Analog Signals use attributes of the medium to convey the signal's information. For example, an aneroid barometer makes use of the angular position of a needle to convey the changes in atmospheric pressure. The signals take any value from a given range and each signal value denotes different information. Each level of the signal signifies a different level of the phenomenon and any change in the signal is meaningful.

It's quite easy to determine if a signal graph is analog or digital. The former one is smooth and continuous while the latter is edge and appears in the form of stepping squares. Given in the image is an analog signal graph representing the change in voltage with the change in time.

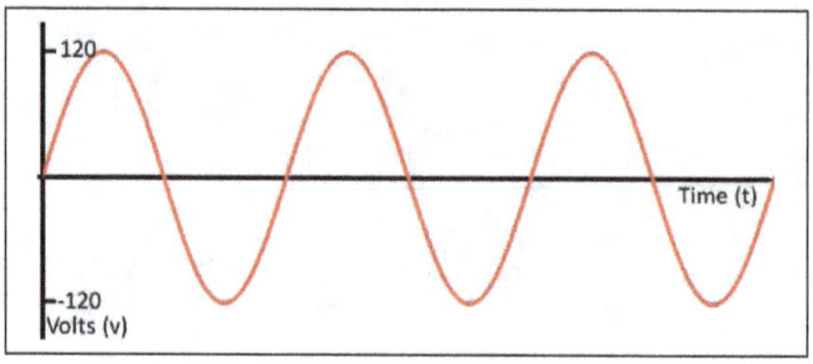

An analog signal graph representing time-voltage graph.

Analog Circuit

Analog circuit elements support a continuous range of voltages, and the information they carry is usually encoded as a time-varying continuous signal similar to those used throughout this text. All of the circuits described thus far are analog circuits. Analog circuitry is a necessary part of most medical instrumentation because the devices that perform physiological measurements, the transducers or "biotransducers," usually produce analog electric signals. This includes devices that measure movement, pressure, bioelectric activity, sound and ultrasound, light, and other forms of electromagnetic energy. Bioelectric signals such as the EEG, ECG, and EMG are analog signals. Before these analog signals can be processed by a digital computer, some type of manipulation is usually required while the signals are still in the analog domain. This analog signal processing may only consist of increasing the amplitude of the signal, but can also include filtering and other basic signal-processing operations. Unlike digital circuitry, the design of analog circuits is often the responsibility of the biomedical engineer. After analog signal processing, the signal is usually converted to a digital signal using an analog-to-digital converter. The components of a typical biomedical instrument are summarized in figure.

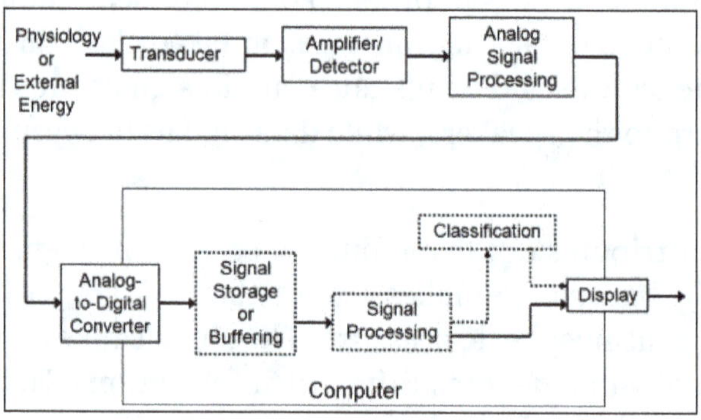

Analogue Filter

Analogue filters are a basic building block of signal processing much used in electronics. Amongst their many applications are the separation of an audio signal before application to bass, mid-range, and tweeter loudspeakers; the combining and later separation of multiple telephone conversations onto a single channel; the selection of a chosen radio station in a radio receiver and rejection of others.

Passive linear electronic analogue filters are those filters which can be described with linear differential equations (linear); they are composed of capacitors, inductors and, sometimes, resistors (passive) and are designed to operate on continuously varying (analogue) signals. There are many linear filters which are not analogue in implementation (digital filter), and there are many electronic filters which may not have a passive topology. Analogue filters are most often used in wave filtering applications, that is, where it is required to pass particular frequency components and to reject others from analogue (continuous-time) signals.

Analogue filters have played an important part in the development of electronics. Especially in the field of telecommunications, filters have been of crucial importance in a number of technological breakthroughs and have been the source of enormous profits for telecommunications companies. It should come as no surprise, therefore, that the early development of filters was intimately connected with transmission lines. Transmission line theory gave rise to filter theory, which initially took a very similar form, and the main application of filters was for use on telecommunication transmission lines. However, the arrival of network synthesis techniques greatly enhanced the degree of control of the designer.

Today, it is often preferred to carry out filtering in the digital domain where complex algorithms are much easier to implement, but analogue filters do still find applications, especially for low-order simple filtering tasks and are often still the norm at higher frequencies where digital technology is still impractical, or at least, less cost effective. Wherever possible, and especially at low frequencies, analogue filters are now implemented in a filter topology which is active in order to avoid the wound components (i.e. inductors, transformers, etc.) required by passive topology.

It is possible to design linear analogue mechanical filters using mechanical components which filter mechanical vibrations or acoustic waves. While there are few applications for such devices in mechanics per se, they can be used in electronics with the addition of transducers to convert to and from the electrical domain. Indeed, some of the earliest ideas for filters were acoustic resonators because the electronics technology was poorly understood at the time. In principle, the design of such filters can be achieved entirely in terms of the electronic counterparts of mechanical quantities, with kinetic energy, potential energy and heat energy corresponding to the energy in inductors, capacitors and resistors respectively.

There are three main stages in the history of passive analogue filter development:

- Simple filters: The frequency dependence of electrical response was known for capacitors and inductors from very early on. The resonance phenomenon was also familiar from an early date and it was possible to produce simple, single-branch filters with these components. Although attempts were made in the 1880s to apply them to telegraphy, these designs proved inadequate for successful frequency-division multiplexing. Network analysis was not yet powerful enough to provide the theory for more complex filters and progress was further hampered by a general failure to understand the frequency domain nature of signals.

- Image filters: Image filter theory grew out of transmission line theory and the design proceeded in a similar manner to transmission line analysis. For the first time filters could be produced that had precisely controllable passbands and other parameters. These developments took place in the 1920s and filters produced to these designs were still in widespread use in the 1980s, only declining as the use of analogue telecommunications has declined. Their immediate application was the economically important development of frequency division multiplexing for use on intercity and international telephony lines.
- Network synthesis filters: The mathematical bases of network synthesis were laid in the 1930s and 1940s. After World War II, network synthesis became the primary tool of filter design. Network synthesis put filter design on a firm mathematical foundation, freeing it from the mathematically sloppy techniques of image design and severing the connection with physical lines. The essence of network synthesis is that it produces a design that will (at least if implemented with ideal components) accurately reproduce the response originally specified in black box terms.

The letters R, L and C are used with their usual meanings to represent resistance, inductance and capacitance, respectively. In particular they are used in combinations, such as LC, to mean, for instance, a network consisting only of inductors and capacitors. Z is used for electrical impedance, any 2-terminal combination of RLC elements and in some sections D is used for the rarely seen quantity elastance, which is the inverse of capacitance.

Resonance

Early filters utilised the phenomenon of resonance to filter signals. Although electrical resonance had been investigated by researchers from a very early stage, it was at first not widely understood by electrical engineers. Consequently, the much more familiar concept of acoustic resonance found its way into filter design ahead of electrical resonance. Resonance can be used to achieve a filtering effect because the resonant device will respond to frequencies at, or near, to the resonant frequency but will not respond to frequencies far from resonance. Hence frequencies far from resonance are filtered out from the output of the device.

Electrical Resonance

Resonance was noticed early on in experiments with the Leyden jar, invented in 1746. The Leyden jar stores electricity due to its capacitance, and is, in fact, an early form of capacitor. When a Leyden jar is discharged by allowing a spark to jump between the electrodes, the discharge is oscillatory. This was not suspected until 1826, when Felix Savary in France, and later Joseph Henry in the US noted that a steel needle placed close to the discharge does not always magnetise in the same direction. They both independently drew the conclusion that there was a transient oscillation dying with time.

Hermann von Helmholtz in 1847 published his important work on conservation of energy in part of which he used those principles to explain why the oscillation dies away, that it is the resistance of the circuit which dissipates the energy of the oscillation on each successive cycle. Helmholtz also noted that there was evidence of oscillation from the electrolysis experiments of William Hyde Wollaston. Wollaston was attempting to decompose water by electric shock but found that both hydrogen and oxygen were present at both electrodes. In normal electrolysis they would separate, one to each electrode.

Helmholtz explained why the oscillation decayed but he had not explained why it occurred in the first place. This was left to Sir William Thomson (Lord Kelvin) who, in 1853, postulated that there was inductance present in the circuit as well as the capacitance of the jar and the resistance of the load. This established the physical basis for the phenomenon - the energy supplied by the jar was partly dissipated in the load but also partly stored in the magnetic field of the inductor.

So far, the investigation had been on the natural frequency of transient oscillation of a resonant circuit resulting from a sudden stimulus. More important from the point of view of filter theory is the behaviour of a resonant circuit when driven by an external AC signal: there is a sudden peak in the circuit's response when the driving signal frequency is at the resonant frequency of the circuit. James Clerk Maxwell heard of the phenomenon from Sir William Grove in 1868 in connection with experiments on dynamos, and was also aware of the earlier work of Henry Wilde in 1866.

Heinrich Hertz experimentally demonstrated the resonance phenomena by building two resonant circuits, one of which was driven by a generator and the other was tunable and only coupled to the first electromagnetically (i.e., no circuit connection). Hertz showed that the response of the second circuit was at a maximum when it was in tune with the first. The diagrams produced by Hertz in this paper were the first published plots of an electrical resonant response.

Acoustic Resonance

It was acoustic resonance that inspired filtering applications, the first of these being a telegraph system known as the "harmonic telegraph". Versions are due to Elisha Gray, Alexander Graham Bell, Ernest Mercadier and others. Its purpose was to simultaneously transmit a number of telegraph messages over the same line and represents an early form of frequency division multiplexing (FDM). FDM requires the sending end to be transmitting at different frequencies for each individual communication channel. This demands individual tuned resonators, as well as filters to separate out the signals at the receiving end. The harmonic telegraph achieved this with electromagnetically driven tuned reeds at the transmitting end which would vibrate similar reeds at the receiving end. Only the reed with the same resonant frequency as the transmitter would vibrate to any appreciable extent at the receiving end.

Incidentally, the harmonic telegraph directly suggested to Bell the idea of the telephone. The reeds can be viewed as transducers converting sound to and from an electrical signal. It is no great leap from this view of the harmonic telegraph to the idea that speech can be converted to and from an electrical signal.

Early Multiplexing

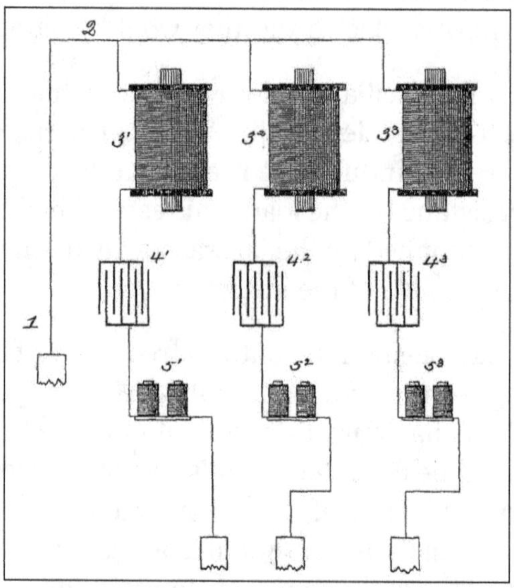

Hutin and Leblanc's multiple telegraph filter showing the use of resonant circuits in filtering.

By the 1890s electrical resonance was much more widely understood and had become a normal part of the engineer's toolkit. In 1891 Hutin and Leblanc patented an FDM scheme for telephone circuits using resonant circuit filters. Rival patents were filed in 1892 by Michael Pupin and John Stone Stone with similar ideas, priority eventually being awarded to Pupin. However, no scheme using just simple resonant circuit filters can successfully multiplex (i.e. combine) the wider bandwidth of telephone channels (as opposed to telegraph) without either an unacceptable restriction of speech bandwidth or a channel spacing so wide as to make the benefits of multiplexing uneconomic.

The basic technical reason for this difficulty is that the frequency response of a simple filter approaches a fall of 6 dB/octave far from the point of resonance. This means that if telephone channels are squeezed in side-by-side into the frequency spectrum, there will be crosstalk from adjacent channels in any given channel. What is required is a much more sophisticated filter that has a flat frequency response in the required passband like a low-Q resonant circuit, but that rapidly falls in response (much faster than 6 dB/octave) at the transition from passband to stopband like a high-Q resonant circuit. Obviously, these are contradictory requirements to be met with a single resonant circuit. The solution to these needs was founded in the theory of transmission lines and consequently the necessary filters did not become available until this theory was fully developed. At this early stage the idea of signal bandwidth, and hence the need for filters to match to it, was not fully understood; indeed, it was as late as 1920 before the concept of bandwidth was fully established. For early radio, the concepts of Q-factor, selectivity and tuning sufficed. This was all to change with the developing theory of transmission lines on which image filters are based.

At the turn of the century as telephone lines became available, it became popular to add telegraph onto telephone lines with an earth return phantom circuit. An LC filter was required to prevent telegraph clicks being heard on the telephone line. From the 1920s onwards, telephone lines, or balanced lines dedicated to the purpose, were used for FDM telegraph at audio frequencies. The first of these systems in the UK was a Siemens and Halske installation between London and Manchester. GEC and AT&T also had FDM systems. Separate pairs were used for the send and receive signals. The Siemens and GEC systems had six channels of telegraph in each direction, the AT&T system had twelve. All of these systems used electronic oscillators to generate a different carrier for each telegraph signal and required a bank of band-pass filters to separate out the multiplexed signal at the receiving end.

Transmission Line Theory

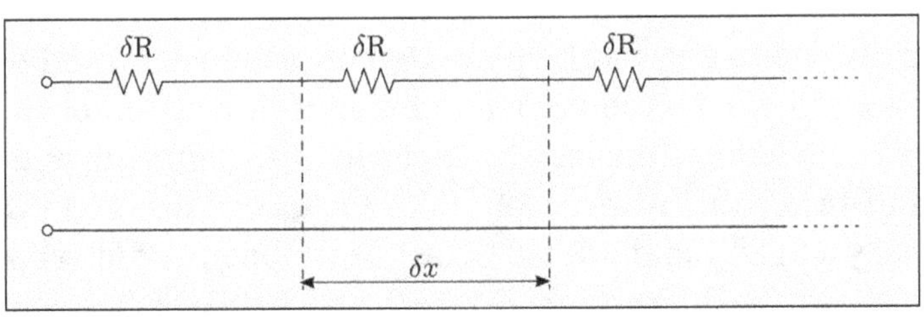

Ohm's model of the transmission line was simply resistance.

The earliest model of the transmission line was probably described by Georg Ohm who established that resistance in a wire is proportional to its length. The Ohm model thus included only resistance. Latimer Clark noted that signals were delayed and elongated along a cable, an undesirable form of distortion now called dispersion but then called retardation, and Michael Faraday established that this was due to the capacitance present in the transmission line. Lord Kelvin found the correct mathematical description needed in his work on early transatlantic cables; he arrived at an equation identical to the conduction of a heat pulse along a metal bar. This model incorporates only resistance and capacitance, but that is all that was needed in undersea cables dominated by capacitance effects. Kelvin's model predicts a limit on the telegraph signalling speed of a cable but Kelvin still did not use the concept of bandwidth, the limit was entirely explained in terms of the dispersion of the telegraph symbols. The mathematical model of the transmission line reached its fullest development with Oliver Heaviside. Heaviside introduced series inductance and shunt conductance into the model making four distributed elements in all. This model is now known as the telegrapher's equation and the distributed-element parameters are called the primary line constants.

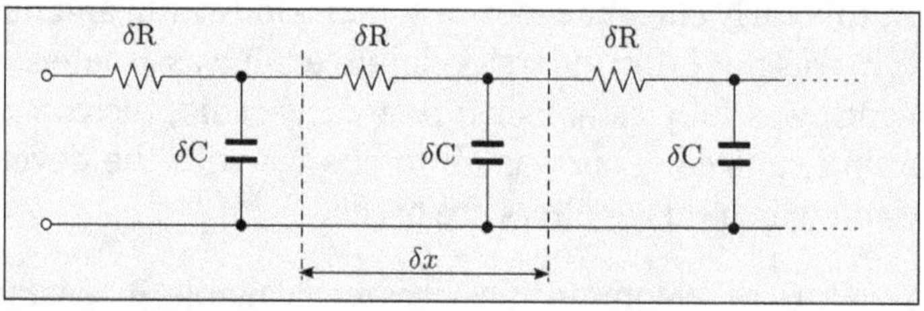

Lord Kelvin's model of the transmission line accounted for capacitance and the dispersion it caused. The diagram represents Kelvin's model translated into modern terms using infinitesimal elements, but this was not the actual approach used by Kelvin.

From the work of Heaviside it had become clear that the performance of telegraph lines, and most especially telephone lines, could be improved by the addition of inductance to the line. George Campbell at AT&T implemented this idea by inserting loading coils at intervals along the line. Campbell found that as well as the desired improvements to the line's characteristics in the passband there was also a definite frequency beyond which signals could not be passed without great attenuation. This was a result of the loading coils and the line capacitance forming a low-pass filter, an effect that is only apparent on lines incorporating lumped components such as the loading coils. This naturally led Campbell to produce a filter with ladder topology, a glance at the circuit diagram of this filter is enough to see its relationship to a loaded transmission line. The cut-off phenomenon is an undesirable side-effect as far as loaded lines are concerned but for telephone FDM filters it is precisely what is required. For this application, Campbell produced band-pass filters to the same ladder topology by replacing the inductors and capacitors with resonators and anti-resonators respectively. Both the loaded line and FDM were of great benefit economically to AT&T and this led to fast development of filtering from this point onwards.

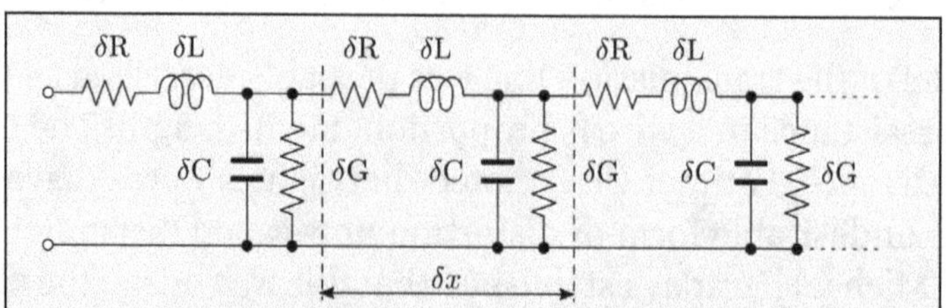

Heaviside's model of the transmission line. L, R, C and G in all three diagrams are the primary line constants. The infinitesimals δL, δR, δC and δG are to be understood as Lδx, Rδx, Cδx and Gδx respectively.

Image Filters

Campbell's sketch of the low-pass version of his filter from his 1915 patent showing the now ubiquitous ladder topology with capacitors for the ladder rungs and inductors for the stiles. Filters of more modern design also often adopt the same ladder topology as used by Campbell. It should be understood that although superficially similar, they are really quite different. The ladder construction is essential to the Campbell filter and all the sections have identical element values. Modern designs can be realised in any number of topologies, choosing the ladder topology is merely a matter of convenience. Their response is quite different (better) than Campbell's and the element values, in general, will all be different.

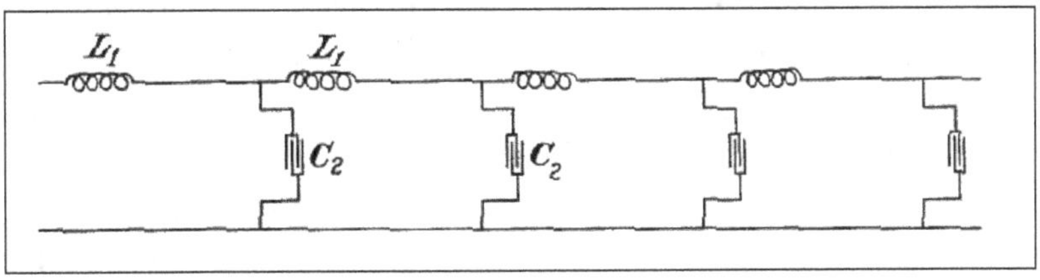

The filters designed by Campbell were named wave filters because of their property of passing some waves and strongly rejecting others. The method by which they were designed was called the image parameter method and filters designed to this method are called image filters. The image method essentially consists of developing the transmission constants of an infinite chain of identical filter sections and then terminating the desired finite number of filter sections in the image impedance. This exactly corresponds to the way the properties of a finite length of transmission line are derived from the theoretical properties of an infinite line, the image impedance corresponding to the characteristic impedance of the line.

From 1920 John Carson, also working for AT&T, began to develop a new way of looking at signals using the operational calculus of Heaviside which in essence is working in the frequency domain. This gave the AT&T engineers a new insight into the way their filters were working and led Otto Zobel to invent many improved forms. Carson and Zobel steadily demolished many of the old ideas. For instance the old telegraph engineers thought of the signal as being a single frequency and this idea persisted into the age of radio with some still believing that frequency modulation (FM) transmission could be achieved with a smaller bandwidth than the baseband signal right up until the publication of Carson's 1922 paper. Another advance concerned the nature of noise, Carson and Zobel treated noise as a random process with a continuous bandwidth, an idea that was well ahead of its time, and thus limited the amount of noise that it was possible to remove by filtering to that part of the noise spectrum which fell outside the passband. This too, was not generally accepted at first, notably being opposed by Edwin Armstrong (who ironically, actually succeeded in reducing noise with wide-band FM) and was only finally settled with the work of Harry Nyquist whose thermal noise power formula is well known today.

Several improvements were made to image filters and their theory of operation by Otto Zobel. Zobel coined the term constant k filter (or k-type filter) to distinguish Campbell's filter from later types, notably Zobel's m-derived filter (or m-type filter). The particular problems Zobel was trying to address with these new forms were impedance matching into the end terminations and improved steepness of roll-off. These were achieved at the cost of an increase in filter circuit complexity.

A more systematic method of producing image filters was introduced by Hendrik Bode, and further developed by several other investigators including Piloty and Wilhelm Cauer. Rather than enumerate the behaviour (transfer function, attenuation function, delay function and so on) of a specific circuit, instead a requirement for the image impedance itself was developed. The image impedance can be expressed in terms of the open-circuit and short-circuit impedances of the filter as $Z_i = \sqrt{Z_o Z_s}$. Since the image impedance must be real in the passbands and imaginary in the stopbands according to image theory, there is a requirement that the poles and zeroes of Z_o and Z_s cancel in the passband and correspond in the stopband. The behaviour of the filter can be entirely defined in terms of the positions in the complex plane of these pairs of poles and zeroes. Any circuit which has the requisite poles and zeroes will also have the requisite response. Cauer pursued two related questions arising from this technique: what specification of poles and zeroes are realisable as passive filters; and what realisations are equivalent to each other. The results of this work led Cauer to develop a new approach, now called network synthesis.

This "poles and zeroes" view of filter design was particularly useful where a bank of filters, each operating at different frequencies, are all connected across the same transmission line. The earlier approach was unable to deal properly with this situation, but the poles and zeroes approach could embrace it by specifying a constant impedance for the combined filter. This problem was originally related to FDM telephony but frequently now arises in loudspeaker crossover filters.

Network Synthesis Filters

The essence of network synthesis is to start with a required filter response and produce a network that delivers that response, or approximates to it within a specified boundary. This is the inverse of network analysis which starts with a given network and by applying the various electric circuit theorems predicts the response of the network. The term was first used with this meaning in the doctoral thesis of Yuk-Wing Lee and apparently arose out of a conversation with Vannevar Bush. The advantage of network synthesis over previous methods is that it provides a solution which precisely meets the design specification. This is not the case with image filters, a degree of experience is required in their design since the image filter only meets the design specification in the unrealistic case of being terminated in its own image impedance, to produce which would require the exact circuit being sought. Network synthesis on the other hand, takes care of the termination impedances simply by incorporating them into the network being designed.

The development of network analysis needed to take place before network synthesis was possible. The theorems of Gustav Kirchhoff and others and the ideas of Charles Steinmetz (phasors) and Arthur Kennelly (complex impedance) laid the groundwork. The concept of a port also played a part in the development of the theory, and proved to be a more useful idea than network terminals. The first milestone on the way to network synthesis was an important paper by Ronald M. Foster, A Reactance Theorem, in which Foster introduces the idea of a driving point impedance, that is, the impedance that is connected to the generator. The expression for this impedance determines the response of the filter and vice versa, and a realisation of the filter can be obtained by expansion of this expression. It is not possible to realise any arbitrary impedance expression as a network. Foster's reactance theorem stipulates necessary and sufficient conditions for realisability: that the reactance must be algebraically increasing with frequency and the poles and zeroes must alternate.

Wilhelm Cauer expanded on the work of Foster and was the first to talk of realisation of a one-port impedance with a prescribed frequency function. Foster's work considered only reactances (i.e., only LC-kind circuits). Cauer generalised this to any 2-element kind one-port network, finding there was an isomorphism between them. He also found ladder realisations of the network using Thomas Stieltjes' continued fraction expansion. This work was the basis on which network synthesis was built, although Cauer's work was not at first used much by engineers, partly because of the intervention of World War II, because Cauer presented his results using topologies that required mutually coupled inductors and ideal transformers. Designers tend to avoid the complication of mutual inductances and transformers where possible, although transformer-coupled double-tuned amplifiers are a common way of widening bandwidth without sacrificing selectivity.

Image Method vs. Synthesis

Image filters continued to be used by designers long after the superior network synthesis techniques were available. Part of the reason for this may have been simply inertia, but it was largely due to the greater computation required for network synthesis filters, often needing a mathematical iterative process. Image filters, in their simplest form, consist of a chain of repeated, identical sections. The design can be improved simply by adding more sections and the computation required to produce the initial section is on the level of "back of an envelope" designing. In the case of network synthesis filters, on the other hand, the filter is designed as a whole, single entity and to add more sections (i.e., increase the order) the designer would have no option but to go back to the beginning and start over. The advantages of synthesised designs are real, but they are not overwhelming compared to what a skilled image designer could achieve, and in many cases it was more cost effective to dispense with time-consuming calculations. This is simply not an issue with the modern availability of computing power, but in the 1950s it was non-existent, in the 1960s and 1970s available only at cost, and not finally becoming widely available to all designers until the 1980s with the advent of the desktop personal computer. Image filters continued to be designed up to that point and many remained in service into the 21st century.

The computational difficulty of the network synthesis method was addressed by tabulating the component values of a prototype filter and then scaling the frequency and impedance and transforming the bandform to those actually required. This kind of approach, or similar, was already in use with image filters, for instance by Zobel, but the concept of a "reference filter" is due to Sidney Darlington. Darlington, was also the first to tabulate values for network synthesis prototype filters, nevertheless it had to wait until the 1950s before the Cauer-Darlington elliptic filter first came into use.

Once computational power was readily available, it became possible to easily design filters to minimise any arbitrary parameter, for example time delay or tolerance to component variation. The difficulties of the image method were firmly put in the past, and even the need for prototypes became largely superfluous. Furthermore, the advent of active filters eased the computation difficulty because sections could be isolated and iterative processes were not then generally necessary.

Realisability and Equivalence

Realisability (that is, which functions are realisable as real impedance networks) and equivalence (which networks equivalently have the same function) are two important questions in network synthesis. Following an analogy with Lagrangian mechanics, Cauer formed the matrix equation,

$$[\mathbf{A}] = s^2[\mathbf{L}] + s[\mathbf{R}] + [\mathbf{D}] = s[\mathbf{Z}]$$

where **[Z]**,**[R]**,**[L]** and **[D]** are the *nxn* matrices of, respectively, impedance, resistance, inductance and elastance of an *n*-mesh network and *s* is the complex frequency operator. Here **[R]**,**[L]** and **[D]** have associated energies corresponding to the kinetic, potential and dissipative heat energies, respectively, in a mechanical system and the already known results from mechanics could be applied here. Cauer determined the driving point impedance by the method of Lagrange multipliers;

$$Z_p(s) = \frac{\det[\mathbf{A}]}{s a_{11}}$$

where a_{11} is the complement of the element A_{11} to which the one-port is to be connected. From stability theory Cauer found that **[R]**, **[L]** and **[D]** must all be positive-definite matrices for $Z_p(s)$ to be realisable if ideal transformers are not excluded. Realisability is only otherwise restricted by practical limitations on topology. This work is also partly due to Otto Brune, who worked with Cauer in the US prior to Cauer returning to Germany. A well known condition for realisability of a one-port rational impedance due to Cauer is that it must be a function of *s* that is analytic in the right halfplane ($\sigma > 0$), have a positive real part in the right halfplane and take on real values on the real axis.

This follows from the Poisson integral representation of these functions. Brune coined the term positive-real for this class of function and proved that it was a necessary and sufficient condition (Cauer had only proved it to be necessary) and they extended the work to LC multiports. A theorem due to Sidney Darlington states that any positive-real function *Z*(*s*) can be realised as a lossless two-port terminated in a positive resistor R. No resistors within the network are necessary to realise the specified response.

As for equivalence, Cauer found that the group of real affine transformations,

$$[\mathbf{T}]^T[\mathbf{A}][\mathbf{T}]$$

where,

$$[\mathbf{T}] = \begin{bmatrix} 1 & 0 \cdots 0 \\ T_{21} & T_{22} \cdots T_{2n} \\ \cdot & \cdots \\ T_{n1} & T_{n2} \cdots T_{nn} \end{bmatrix}$$

is invariant in $Z_p(s)$, that is, all the transformed networks are equivalents of the original.

Approximation

The approximation problem in network synthesis is to find functions which will produce realisable networks approximating to a prescribed function of frequency within limits arbitrarily set. The approximation problem is an important issue since the ideal function of frequency required will commonly be unachievable with rational networks. For instance, the ideal prescribed function is often taken to be the unachievable lossless transmission in the passband, infinite attenuation in the stopband and a vertical transition between the two. However, the ideal function can be approximated with a rational function, becoming ever closer to the ideal the higher the order of the polynomial. The first to address this problem was Stephen Butterworth using his Butterworth polynomials. Independently, Cauer used Chebyshev polynomials, initially applied to image filters, and not to the now well-known ladder realisation of this filter.

Butterworth Filter

Butterworth filters are an important class of filters due to Stephen Butterworth which are now recognised as being a special case of Cauer's elliptic filters. Butterworth discovered this filter independently of Cauer's work and implemented it in his version with each section isolated from the next with a valve amplifier which made calculation of component values easy since the filter sections could not interact with each other and each section represented one term in the Butterworth polynomials. This gives Butterworth the credit for being both the first to deviate from image parameter theory and the first to design active filters. It was later shown that Butterworth filters could be implemented in ladder topology without the need for amplifiers. Possibly the first to do so was William Bennett in a patent which presents formulae for component values identical to the modern ones. Bennett, at this stage though, is still discussing the design as an artificial transmission line and so is adopting an image parameter approach despite having produced what would now be considered a network synthesis design. He also does not appear to be aware of the work of Butterworth or the connection between them.

Insertion-loss Method

The insertion-loss method of designing filters is, in essence, to prescribe a desired function of frequency for the filter as an attenuation of the signal when the filter is inserted between the terminations relative to the level that would have been received were the terminations connected to each other via an ideal transformer perfectly matching them. Versions of this theory are due to Sidney Darlington, Wilhelm Cauer and others all working more or less independently and is often taken as synonymous with network synthesis. Butterworth's filter implementation is, in those terms, an insertion-loss filter, but it is a relatively trivial one mathematically since the active amplifiers used by Butterworth ensured that each stage individually worked into a resistive load. Butterworth's filter becomes a non-trivial example when it is implemented entirely with passive components. An even earlier filter which influenced the insertion-loss method was Norton's dual-band filter where the input of two filters are connected in parallel and designed so that the combined input presents a constant resistance. Norton's design method, together with Cauer's canonical LC networks and Darlington's theorem that only LC components were required in the body of the filter resulted in the insertion-loss method. However, ladder topology proved to be more practical than Cauer's canonical forms.

Darlington's insertion-loss method is a generalisation of the procedure used by Norton. In Norton's filter it can be shown that each filter is equivalent to a separate filter unterminated at the common end. Darlington's method applies to the more straightforward and general case of a 2-port LC network terminated at both ends. The procedure consists of the following steps:

- Determine the poles of the prescribed insertion-loss function.
- From that find the complex transmission function.
- From that find the complex reflection coefficients at the terminating resistors.
- Find the driving point impedance from the short-circuit and open-circuit impedances.
- Expand the driving point impedance into an LC (usually ladder) network.

Darlington additionally used a transformation found by Hendrik Bode that predicted the response of a filter using non-ideal components but all with the same Q. Darlington used this transformation in reverse to produce filters with a prescribed insertion-loss with non-ideal components. Such filters have the ideal insertion-loss response plus a flat attenuation across all frequencies.

Elliptic Filters

Elliptic filters are filters produced by the insertion-loss method which use elliptic rational functions in their transfer function as an approximation to the ideal filter response and the result is called a Chebyshev approximation. This is the same Chebyshev approximation technique used by Cauer on image filters but follows the Darlington insertion-loss design method and uses slightly different elliptic functions. Cauer had some contact with Darlington and Bell Labs before WWII (for a time he worked in the US) but during the war they worked independently, in some cases making the same discoveries. Cauer had disclosed the Chebyshev approximation to Bell Labs but had not left them with the proof. Sergei Schelkunoff provided this and a generalisation to all equal ripple problems. Elliptic filters are a general class of filter which incorporate several other important classes as special cases: Cauer filter (equal ripple in passband and stopband), Chebyshev filter (ripple only in passband), reverse Chebyshev filter (ripple only in stopband) and Butterworth filter (no ripple in either band).

Generally, for insertion-loss filters where the transmission zeroes and infinite losses are all on the real axis of the complex frequency plane (which they usually are for minimum component count), the insertion-loss function can be written as;

$$\frac{1}{1+JF^2}$$

where F is either an even (resulting in an antimetric filter) or an odd (resulting in an symmetric filter) function of frequency. Zeroes of F correspond to zero loss and the poles of F correspond to transmission zeroes. J sets the passband ripple height and the stopband loss and these two design requirements can be interchanged. The zeroes and poles of F and J can be set arbitrarily. The nature of F determines the class of the filter;

- If F is a Chebyshev approximation the result is a Chebyshev filter.
- If F is a maximally flat approximation the result is a passband maximally flat filter.
- If $1/F$ is a Chebyshev approximation the result is a reverse Chebyshev filter.
- If $1/F$ is a maximally flat approximation the result is a stopband maximally flat filter.

A Chebyshev response simultaneously in the passband and stopband is possible, such as Cauer's equal ripple elliptic filter.

Other Methods

Darlington considers the topology of coupled tuned circuits to involve a separate approximation technique to the insertion-loss method, but also producing nominally flat passbands and high attenuation stopbands. The most common topology for these is shunt anti-resonators coupled by series capacitors, less commonly, by inductors, or in the case of a two-section filter, by mutual inductance. These are most useful where the design requirement is not too stringent, that is, moderate bandwidth, roll-off and passband ripple.

Other Notable Developments and Applications

Mechanical Filters

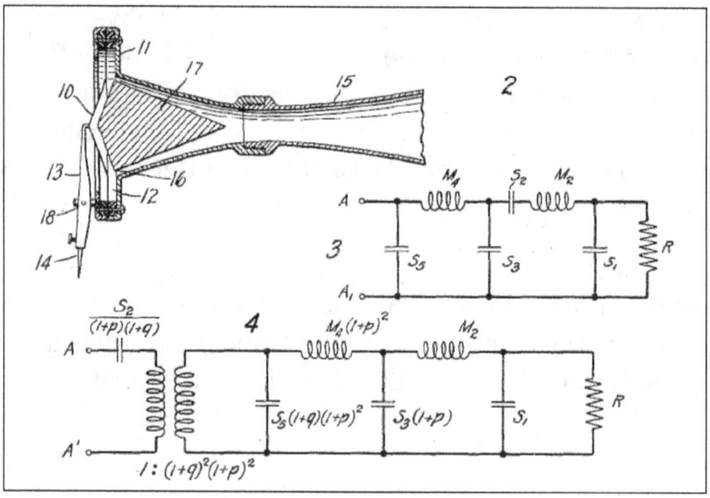

Norton's mechanical filter together with its electrical equivalent circuit. Two equivalents are shown, figure directly corresponds to the physical relationship of the mechanical components; figure is an equivalent transformed circuit arrived at by repeated application of a well known transform, the purpose being to remove the series resonant circuit from the body of the filter leaving a simple *LC* ladder network.

Edward Norton, around 1930, designed a mechanical filter for use on phonograph recorders and players. Norton designed the filter in the electrical domain and then used the correspondence of mechanical quantities to electrical quantities to realise the filter using mechanical components. Mass corresponds to inductance, stiffness to elastance and damping to resistance. The filter was designed to have a maximally flat frequency response.

In modern designs it is common to use quartz crystal filters, especially for narrowband filtering applications. The signal exists as a mechanical acoustic wave while it is in the crystal and is converted by transducers between the electrical and mechanical domains at the terminals of the crystal.

Distributed-element Filters

Distributed-element filters are composed of lengths of transmission line that are at least a significant fraction of a wavelength long. The earliest non-electrical filters were all of this type. William Herschel, for instance, constructed an apparatus with two tubes of different lengths which attenuated some frequencies but not others. Joseph-Louis Lagrange studied waves on a string periodically loaded with weights. The device was never studied or used as a filter by either Lagrange or later investigators such as Charles Godfrey. However, Campbell used Godfrey's results by analogy to calculate the number of loading coils needed on his loaded lines, the device that led to his electrical filter development. Lagrange, Godfrey, and Campbell all made simplifying assumptions in their calculations that ignored the distributed nature of their apparatus. Consequently, their models did not show the multiple passbands that are a characteristic of all distributed-element filters. The first electrical filters that were truly designed by distributed-element principles are due to Warren P. Mason starting in 1927.

Transversal Filters

Transversal filters are not usually associated with passive implementations but the concept can be found in a Wiener and Lee patent from 1935 which describes a filter consisting of a cascade of all-pass sections. The outputs of the various sections are summed in the proportions needed to result in the required frequency function. This works by the principle that certain frequencies will be in, or close to antiphase, at different sections and will tend to cancel when added. These are the frequencies rejected by the filter and can produce filters with very sharp cut-offs. This approach did not find any immediate applications, and is not common in passive filters. However, the principle finds many applications as an active delay line implementation for wide band discrete-time filter applications such as television, radar and high-speed data transmission.

Matched Filter

The purpose of matched filters is to maximise the signal-to-noise ratio (S/N) at the expense of pulse shape. Pulse shape, unlike many other applications, is unimportant in radar while S/N is the primary limitation on performance. The filters were introduced during WWII by Dwight North and are often eponymously referred to as "North filters".

Filters for Control Systems

Control systems have a need for smoothing filters in their feedback loops with criteria to maximise the speed of movement of a mechanical system to the prescribed mark and at the same time minimise overshoot and noise induced motions. A key problem here is the extraction of Gaussian signals from a noisy background. An early paper on this was published during WWII by Norbert Wiener with the specific application to anti-aircraft fire control analogue computers. Rudy Kalman (Kalman filter) later reformulated this in terms of state-space smoothing and prediction where it is known as the linear-quadratic-Gaussian control problem. Kalman started an interest in state-space solutions, but according to Darlington this approach can also be found in the work of Heaviside and earlier.

Modern Practice

LC filters at low frequencies become awkward; the components, especially the inductors, become expensive, bulky, heavy, and non-ideal. Practical 1 H inductors require many turns on a high-permeability core; that material will have high losses and stability issues (e.g., a large temperature coefficient). For applications such as a mains filters, the awkwardness must be tolerated. For low-level, low-frequency, applications, RC filters are possible, but they cannot implement filters with complex poles or zeros. If the application can use power, then amplifiers can be used to make RC active filters that can have complex poles and zeros. In the 1950s, Sallen–Key active RC filters were made with vacuum tube amplifiers; these filters replaced the bulky inductors with bulky and hot vacuum tubes. Transistors offered more power-efficient active filter designs. Later, inexpensive operational amplifiers enabled other active RC filter design topologies. Although active filter designs were commonplace at low frequencies, they were impractical at high frequencies where the amplifiers were not ideal; LC (and transmission line) filters were still used at radio frequencies.

Gradually, the low frequency active RC filter was supplanted by the switched-capacitor filter that operated in the discrete time domain rather than the continuous time domain. All of these filter technologies require precision components for high performance filtering, and that often requires that the filters be tuned. Adjustable components are expensive, and the labor to do the tuning can be significant. Tuning the poles and zeros of a 7th-order elliptic filter is not a simple exercise. Integrated circuits have made digital computation inexpensive, so now low frequency filtering is done with digital signal processors. Such digital filters have no problem implementing ultra-precise (and stable) values, so no tuning or adjustment is required. Digital filters also don't have to worry about stray coupling paths and shielding the individual filter sections from one another. One downside is the digital signal processing may consume much more power than an equivalent LC filter. Inexpensive digital technology has largely supplanted analogue implementations of filters. However, there is still an occasional place for them in the simpler applications such as coupling where sophisticated functions of frequency are not needed. Passive filters are still the technology of choice at microwave frequencies.

Bipolar Junction Transistor

A bipolar junction transistor (bipolar transistor or BJT) is a type of transistor that uses both electrons and holes as charge carriers.

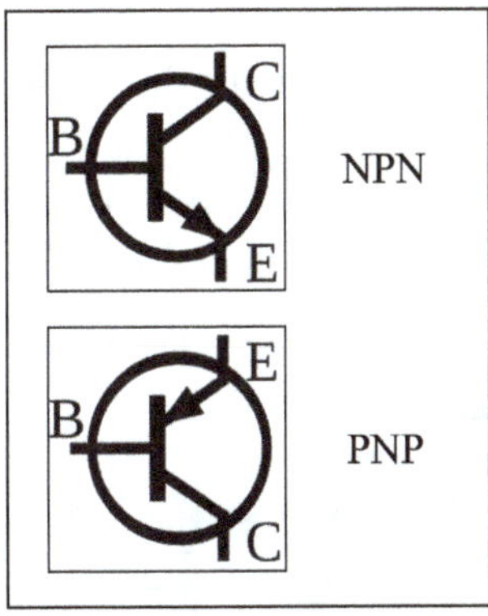

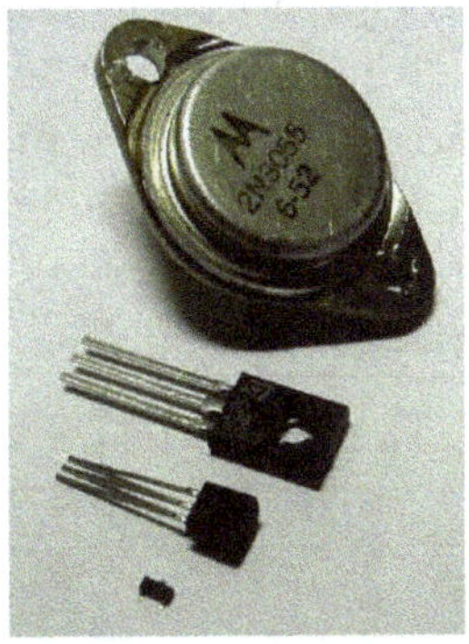

BJT schematic symbols. Typical individual BJT packages.

Unipolar transistors, such as field-effect transistors, use only one kind of charge carrier. BJTs use two junctions between two semiconductor types, n-type and p-type. BJTs are manufactured in two types: NPN and PNP, and are available as individual components, or fabricated in integrated circuits, often in large numbers.

Usage

BJTs can be used as amplifiers or switches. This ability gives them many applications in electronic equipment such as computers, televisions, mobile phones, audio amplifiers, industrial control, and radio transmitters.

Current Direction Conventions

By convention, the direction of current on diagrams is shown as the direction that a positive charge would move. This is called conventional current. However, current in many metal conductors is due to the flow of electrons. Because electrons carry a negative charge, they move in the direction opposite to conventional current. On the other hand, inside a bipolar transistor, currents can be composed of both positively charged holes and negatively charged electrons. Current arrows are shown in the conventional direction, but labels for the movement of holes and electrons show their actual direction inside the transistor. The arrow on the symbol for bipolar transistors indicates the PN junction between base and emitter and points in the direction in which conventional current travels.

Function

BJTs are available in two types, or polarities, known as PNP and NPN based on the doping types of the three main terminal regions. An NPN transistor comprises two semiconductor junctions that share a thin p-doped region, and a PNP transistor comprises two semiconductor junctions that share a thin n-doped region.

Charge flow in a BJT is due to diffusion of charge carriers across a junction between two regions of different charge concentrations. The regions of a BJT are called emitter, base, and collector. A discrete transistor has three leads for connection to these regions.

Typically, the emitter region is heavily doped compared to the other two layers, and the collector is doped much lighter than the base (collector doping is typically ten times lighter than base doping). By design, most of the BJT collector current is due to the flow of charge carriers (electrons or holes) injected from a heavily doped emitter into the base where they are minority carriers that diffuse toward the collector, and so BJTs are classified as minority-carrier devices.

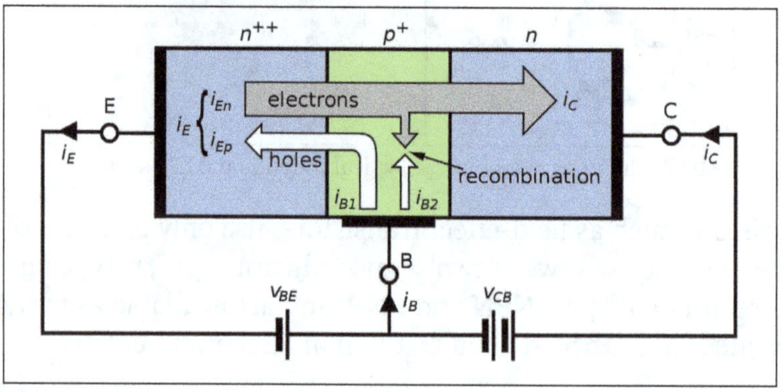

NPN BJT with forward-biased E–B junction and reverse-biased B–C junction.

In typical operation of an NPN device, the base–emitter junction is forward-biased, which means that the p-doped side of the junction is at a more positive potential than the n-doped side, and the base–collector junction is reverse-biased. When forward bias is applied to the base–emitter junction, the equilibrium between the thermally generated carriers and the repelling electric field of the n-doped emitter depletion region is disturbed. This allows thermally excited electrons to inject from the emitter into the base region. These electrons diffuse through the base from the region of high concentration near the emitter toward the region of low concentration near the collector. The electrons in the base are called minority carriers because the base is doped p-type, which makes holes the majority carrier in the base. In a PNP device, analogous behaviour occurs, but with holes as the dominant current carriers.

To minimize the fraction of carriers that recombine before reaching the collector–base junction, the transistor's base region must be thin enough that carriers can diffuse across it in much less time than the semiconductor's minority-carrier lifetime. Having a lightly doped base ensures recombination rates are low. In particular, the thickness of the base must be much less than the diffusion length of the electrons. The collector–base junction is reverse-biased, and so negligible electron injection occurs from the collector to the base, but carriers that are injected into the base and diffuse to reach the collector-base depletion region are swept into the collector by the electric field in the depletion region. The thin shared base and asymmetric collector–emitter doping are what differentiates a bipolar transistor from two separate and oppositely biased diodes connected in series.

Voltage, Current and Charge Control

The collector–emitter current can be viewed as being controlled by the base–emitter current (current control), or by the base–emitter voltage (voltage control). These views are related by the current–voltage relation of the base–emitter junction, which is the usual exponential current–voltage curve of a p–n junction (diode).

The explanation for collector current is the concentration gradient of minority carriers in the base region. Due to low-level injection (in which there are much fewer excess carriers than normal majority carriers) the ambipolar transport rates (in which the excess majority and minority carriers flow at the same rate) is in effect determined by the excess minority carriers.

Detailed transistor models of transistor action, such as the Gummel–Poon model, account for the distribution of this charge explicitly to explain transistor behaviour more exactly. The charge-control view easily handles phototransistors, where minority carriers in the base region are created by the absorption of photons, and handles the dynamics of turn-off, or recovery time, which depends on charge in the base region recombining. However, because base charge is not a signal that is visible at the terminals, the current- and voltage-control views are generally used in circuit design and analysis.

In analog circuit design, the current-control view is sometimes used because it is approximately linear. That is, the collector current is approximately β_F times the base current. Some basic circuits can be designed by assuming that the base-emitter voltage is approximately constant and that collector current is β times the base current. However, to accurately and reliably design production BJT circuits, the voltage-control (for example, Ebers–Moll) model is required. The voltage-control model requires an exponential function to be taken into account, but when it is linearized such that the transistor can be modeled as a transconductance, as in the Ebers–Moll model, design for circuits such as differential amplifiers again becomes a mostly linear problem, so the voltage-control view is often preferred. For translinear circuits, in which the exponential I–V curve is key to the operation, the transistors are usually modeled as voltage-controlled current sources whose transconductance is proportional to their collector current. In general, transistor-level circuit analysis is performed using SPICE or a comparable analog-circuit simulator, so mathematical model complexity is usually not of much concern to the designer, but a simplified view of the characteristics allows designs to be created following a logical process.

Turn-on, Turn-off and Storage Delay

Bipolar transistors, and particularly power transistors, have long base-storage times when they are driven into saturation; the base storage limits turn-off time in switching applications. A Baker clamp can prevent the transistor from heavily saturating, which reduces the amount of charge stored in the base and thus improves switching time.

Transistor Characteristics: Alpha (α) and Beta (β)

The proportion of carriers able to cross the base and reach the collector is a measure of the BJT efficiency. The heavy doping of the emitter region and light doping of the base region causes many more electrons to be injected from the emitter into the base than holes to be injected from the base into the emitter. A thin and lightly-doped base region means that most of the minority carriers that are injected into the base will diffuse to the collector and not recombine.

The common-emitter current gain is represented by β_F or the h-parameter h_{FE}; it is approximately the ratio of the DC collector current to the DC base current in forward-active region. It is typically greater than 50 for small-signal transistors, but can be smaller in transistors designed for high-power applications. Both injection efficiency and recombination in the base reduce the BJT gain.

Another useful characteristic is the common-base current gain, α_F. The common-base current gain is approximately the gain of current from emitter to collector in the forward-active region. This ratio usually has a value close to unity; between 0.980 and 0.998. It is less than unity due to recombination of charge carriers as they cross the base region.

Alpha and beta are related by the following identities:

$$\alpha_F = \frac{I_C}{I_E}, \qquad \beta_F = \frac{I_C}{I_B},$$

$$\alpha_F = \frac{\beta_F}{1+\beta_F} \qquad \Leftrightarrow \beta_F = \frac{\alpha_F}{1-\alpha_F}.$$

Beta is a convenient figure of merit to describe the performance of a bipolar transistor, but is not a fundamental physical property of the device. Bipolar transistors can be considered voltage-controlled devices (fundamentally the collector current is controlled by the base-emitter voltage; the base current could be considered a defect and is controlled by the characteristics of the base-emitter junction and recombination in the base). In many designs beta is assumed high enough so that base current has a negligible effect on the circuit. In some circuits (generally switching circuits), sufficient base current is supplied so that even the lowest beta value a particular device may have will still allow the required collector current to flow.

Structure

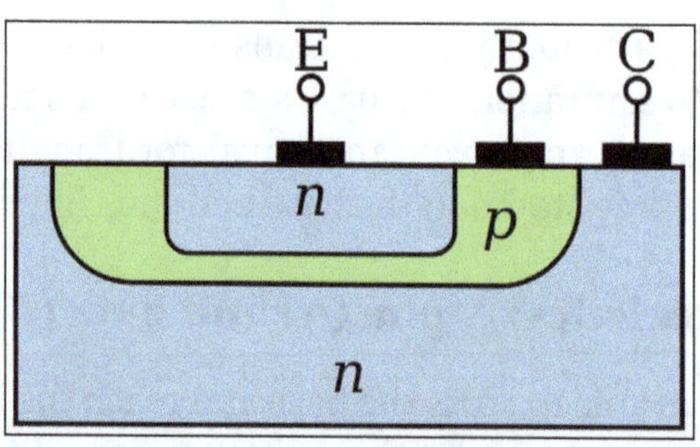

Simplified cross section of a planar *NPN* bipolar junction transistor.

A BJT consists of three differently doped semiconductor regions: the emitter region, the base region and the collector region. These regions are, respectively, p type, n type and p type in a PNP transistor, and n type, p type and n type in an NPN transistor. Each semiconductor region is connected to a terminal, appropriately labeled: emitter (E), base (B) and collector (C).

The base is physically located between the emitter and the collector and is made from lightly doped, high-resistivity material. The collector surrounds the emitter region, making it almost impossible for the electrons injected into the base region to escape without being collected, thus making the resulting value of α very close to unity, and so, giving the transistor a large β. A cross-section view of a BJT indicates that the collector–base junction has a much larger area than the emitter–base junction.

The bipolar junction transistor, unlike other transistors, is usually not a symmetrical device. This means that interchanging the collector and the emitter makes the transistor leave the forward active mode and start to operate in reverse mode. Because the transistor's internal structure is usually optimized for forward-mode operation, interchanging the collector and the emitter makes the values of α and β in reverse operation much smaller than those in forward operation; often the α of the reverse mode is lower than 0.5. The lack of symmetry is primarily due to the doping ratios of the emitter and the collector. The emitter is heavily doped, while the collector is lightly doped, allowing a large reverse bias voltage to be applied before the collector–base junction breaks down. The collector–base junction is reverse biased in normal operation. The reason the emitter is heavily doped is to increase the emitter injection efficiency: the ratio of carriers injected by the emitter to those injected by the base. For high current gain, most of the carriers injected into the emitter–base junction must come from the emitter.

Die of a KSY34 high-frequency NPN transistor.
Bond wires connect to the base and emitter.

The low-performance "lateral" bipolar transistors sometimes used in CMOS processes are sometimes designed symmetrically, that is, with no difference between forward and backward operation.

Small changes in the voltage applied across the base–emitter terminals cause the current between the emitter and the collector to change significantly. This effect can be used to amplify the input voltage or current. BJTs can be thought of as voltage-controlled current sources, but are more simply characterized as current-controlled current sources, or current amplifiers, due to the low impedance at the base.

Early transistors were made from germanium but most modern BJTs are made from silicon. A significant minority are also now made from gallium arsenide, especially for very high speed applications.

NPN

The symbol of an NPN BJT. A mnemonic for the symbol is "not pointing in".

NPN is one of the two types of bipolar transistors, consisting of a layer of P-doped semiconductor (the "base") between two N-doped layers. A small current entering the base is amplified to produce a large collector and emitter current. That is, when there is a positive potential difference measured from the base of an NPN transistor to its emitter (that is, when the base is high relative to the emitter), as well as a positive potential difference measured from the collector to the emitter, the transistor becomes active. In this "on" state, current flows from the collector to the emitter of the transistor. Most of the current is carried by electrons moving from emitter to collector as minority carriers in the P-type base region. To allow for greater current and faster operation, most bipolar transistors used today are NPN because electron mobility is higher than hole mobility.

PNP

The symbol of a PNP BJT. A mnemonic for the symbol is "points in proudly".

The other type of BJT is the PNP, consisting of a layer of N-doped semiconductor between two layers of P-doped material. A small current leaving the base is amplified in the collector output. That is, a PNP transistor is "on" when its base is pulled **low** relative to the emitter. In a PNP transistor, the emitter–base region is forward biased, so holes are injected into the base as minority carriers. The base is very thin, and most of the holes cross the reverse-biased base–collector junction to the collector.

The arrows in the NPN and PNP transistor symbols indicate the PN junction between the base and emitter. When the device is in forward active or forward saturated mode, the arrow, placed on the emitter leg, points in the direction of the conventional current.

Heterojunction Bipolar Transistor

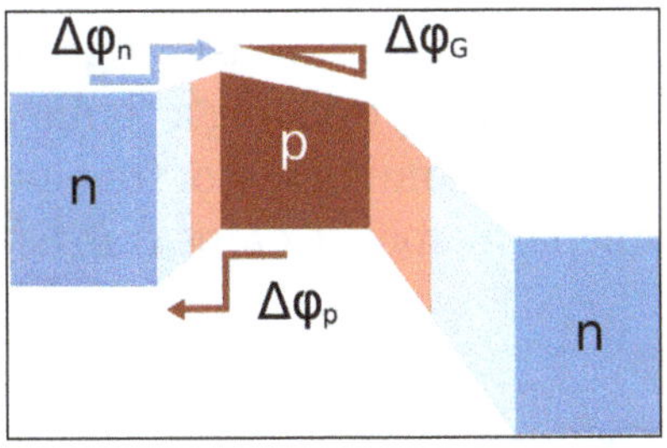

Bands in graded heterojunction NPN bipolar transistor. Barriers indicated for electrons to move from emitter to base and for holes to be injected backward from base to emitter; also, grading of bandgap in base assists electron transport in base region. Light colors indicate depleted regions.

The heterojunction bipolar transistor (HBT) is an improvement of the BJT that can handle signals of very high frequencies up to several hundred GHz. It is common in modern ultrafast circuits, mostly RF systems.

Heterojunction transistors have different semiconductors for the elements of the transistor. Usually the emitter is composed of a larger bandgap material than the base. The figure shows that this difference in bandgap allows the barrier for holes to inject backward from the base into the emitter, denoted in the figure as $\Delta\varphi_p$, to be made large, while the barrier for electrons to inject into the base $\Delta\varphi_n$ is made low. This barrier arrangement helps reduce minority carrier injection from the base when the emitter-base junction is under forward bias, and thus reduces base current and increases emitter injection efficiency.

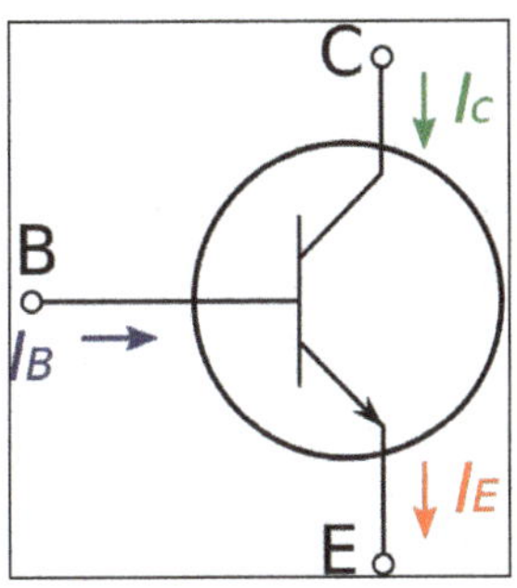

Symbol for NPN bipolar transistor with current flow direction.

The improved injection of carriers into the base allows the base to have a higher doping level, resulting in lower resistance to access the base electrode. In the more traditional

BJT, also referred to as homojunction BJT, the efficiency of carrier injection from the emitter to the base is primarily determined by the doping ratio between the emitter and base, which means the base must be lightly doped to obtain high injection efficiency, making its resistance relatively high. In addition, higher doping in the base can improve figures of merit like the Early voltage by lessening base narrowing.

The grading of composition in the base, for example, by progressively increasing the amount of germanium in a SiGe transistor, causes a gradient in bandgap in the neutral base, denoted in the figure by $\Delta\varphi_G$, providing a "built-in" field that assists electron transport across the base. That drift component of transport aids the normal diffusive transport, increasing the frequency response of the transistor by shortening the transit time across the base.

Two commonly used HBTs are silicon–germanium and aluminum gallium arsenide, though a wide variety of semiconductors may be used for the HBT structure. HBT structures are usually grown by epitaxy techniques like MOCVD and MBE.

Regions of Operation

Junction type	Applied voltages	Junction bias		Mode
		B-E	B-C	
NPN	E < B < C	Forward	Reverse	Forward-active
	E < B > C	Forward	Forward	Saturation
	E > B < C	Reverse	Reverse	Cut-off
	E > B > C	Reverse	Forward	Reverse-active
PNP	E < B < C	Reverse	Forward	Reverse-active
	E < B > C	Reverse	Reverse	Cut-off
	E > B < C	Forward	Forward	Saturation
	E > B > C	Forward	Reverse	Forward-active

Bipolar transistors have four distinct regions of operation, defined by BJT junction biases:

- Forward-active (or simply active): The base–emitter junction is forward biased and the base–collector junction is reverse biased. Most bipolar transistors are designed to afford the greatest common-emitter current gain, β_F, in forward-active mode. If this is the case, the collector–emitter current is approximately proportional to the base current, but many times larger, for small base current variations.
- Reverse-active (or inverse-active or inverted): By reversing the biasing conditions of the forward-active region, a bipolar transistor goes into reverse-active mode. In this mode, the emitter and collector regions switch roles. Because most BJTs are designed to maximize current gain in forward-active mode, the β_F in inverted mode is several times smaller (2–3 times for the ordinary

germanium transistor). This transistor mode is seldom used, usually being considered only for failsafe conditions and some types of bipolar logic. The reverse bias breakdown voltage to the base may be an order of magnitude lower in this region.

- Saturation: With both junctions forward-biased, a BJT is in saturation mode and facilitates high current conduction from the emitter to the collector (or the other direction in the case of NPN, with negatively charged carriers flowing from emitter to collector). This mode corresponds to a logical "on", or a closed switch.

- Cut-off: In cut-off, biasing conditions opposite of saturation (both junctions reverse biased) are present. There is very little current, which corresponds to a logical "off", or an open switch.

Avalanche Breakdown Region

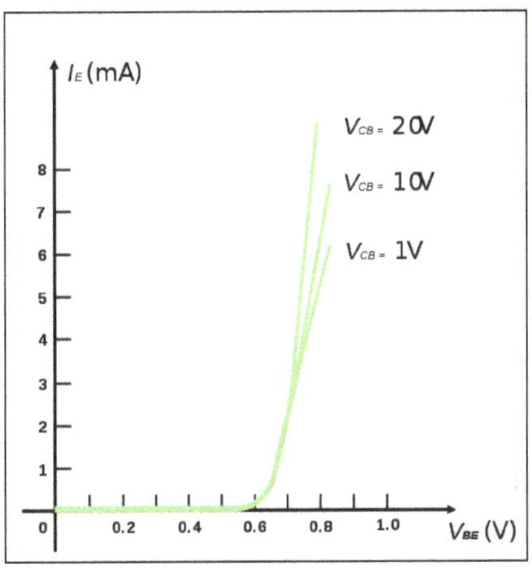

Input characteristics.

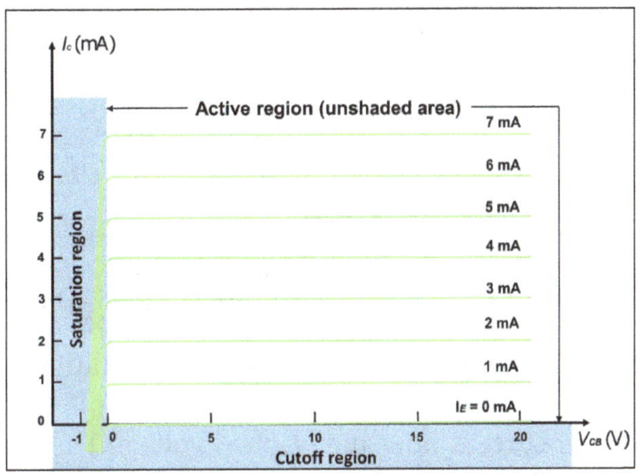

Output characteristics.

The modes of operation can be described in terms of the applied voltages (this description applies to NPN transistors; polarities are reversed for PNP transistors):

- Forward-active: Base higher than emitter, collector higher than base (in this mode the collector current is proportional to base current by β_F).

- Saturation: Base higher than emitter, but collector is not higher than base.

- Cut-off: Base lower than emitter, but collector is higher than base. It means the transistor is not letting conventional current go through from collector to emitter.

- Reverse-active: Base lower than emitter, collector lower than base: reverse conventional current goes through transistor.

In terms of junction biasing: (reverse biased base–collector junction means $V_{bc} < 0$ for NPN, opposite for PNP).

Although these regions are well defined for sufficiently large applied voltage, they overlap somewhat for small (less than a few hundred millivolts) biases. For example, in the typical grounded-emitter configuration of an NPN BJT used as a pulldown switch in digital logic, the "off" state never involves a reverse-biased junction because the base voltage never goes below ground; nevertheless the forward bias is close enough to zero that essentially no current flows, so this end of the forward active region can be regarded as the cutoff region.

Active-mode Transistors in Circuits

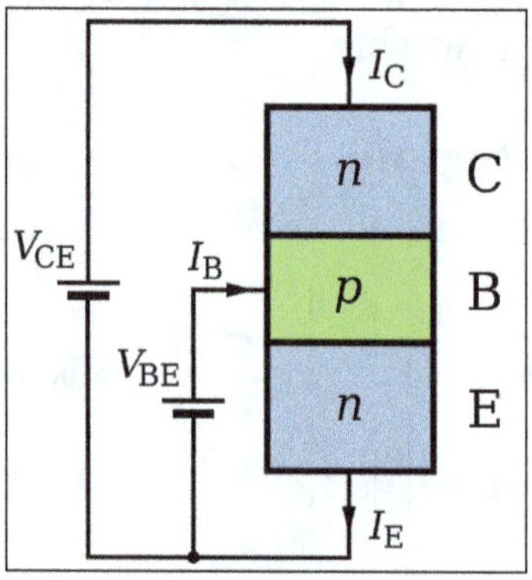

Structure and use of NPN transistor.
Arrow according to schematic.

The diagram shows a schematic representation of an NPN transistor connected to two voltage sources. (The same description applies to a PNP transistor with reversed directions of current flow and applied voltage.) This applied voltage causes the lower P-N junction to become forward biased, allowing a flow of electrons from the emitter into the base. In active mode, the electric field existing between base and collector (caused by V_{CE}) will cause the majority of these electrons to cross the upper P-N junction into the collector to form the collector current I_C. The remainder of the electrons recombine with holes, the majority carriers in the base, making a current through the base connection to form the base current, I_B. The emitter current, I_E, is the total transistor current, which is the sum of the other terminal currents, (i.e., $I_E = I_B + I_C$).

In the diagram, the arrows representing current point in the direction of conventional current – the flow of electrons is in the opposite direction of the arrows because electrons carry negative electric charge. In active mode, the ratio of the collector current to the base current is called the DC current gain. This gain is usually 100 or more, but robust circuit designs do not depend on the exact value. The value of this gain for DC signals is referred to as h_{FE}, and the value of this gain for small signals is referred to as h_{fe}. That is, when a small change in the currents occurs, and sufficient time has passed for the new condition to reach a steady state h_{fe} is the ratio of the change in collector current to the change in base current. The symbol β is used for both h_{FE} and h_{fe}.

The emitter current is related to V_{BE} exponentially. At room temperature, an increase in by approximately 60 mV increases the emitter current by a factor of 10. Because the base current is approximately proportional to the collector and emitter currents, they vary in the same way.

Theory and Modeling

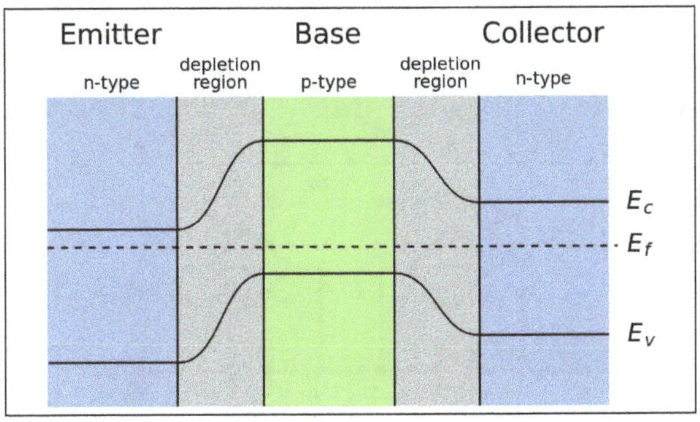

Band diagram for NPN transistor at equilibrium.

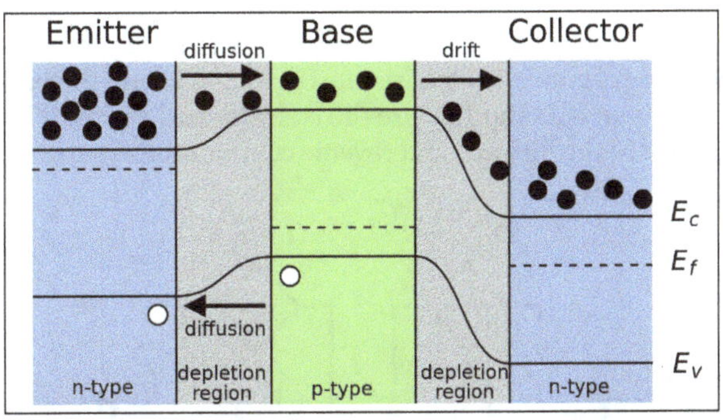

Band diagram for NPN transistor in active mode, showing injection of electrons from emitter to base, and their overshoot into the collector.

Transistors can be thought of as two diodes (P–N junctions) sharing a common region that minority carriers can move through. A PNP BJT will function like two diodes that share an N-type cathode region, and the NPN like two diodes sharing a P-type anode region. Connecting two diodes with wires will not make a transistor, since minority carriers will not be able to get from one P–N junction to the other through the wire.

Both types of BJT function by letting a small current input to the base control an amplified output from the collector. The result is that the transistor makes a good switch that is controlled by its base input. The BJT also makes a good amplifier, since it can multiply a weak input signal to about 100 times its original strength. Networks of transistors are used to make powerful amplifiers with many different applications. In the

NPN transistor in what is called active mode, the base–emitter voltage V_{BE} and collector–base voltage V_{CB} are positive, forward biasing the emitter–base junction and reverse-biasing the collector–base junction. In the active mode of operation, electrons are injected from the forward biased n-type emitter region into the p-type base where they diffuse as minority carriers to the reverse-biased n-type collector and are swept away by the electric field in the reverse-biased collector–base junction.

Large-signal Models

In 1954, Jewell James Ebers and John L. Moll introduced their mathematical model of transistor currents.

Ebers–Moll model

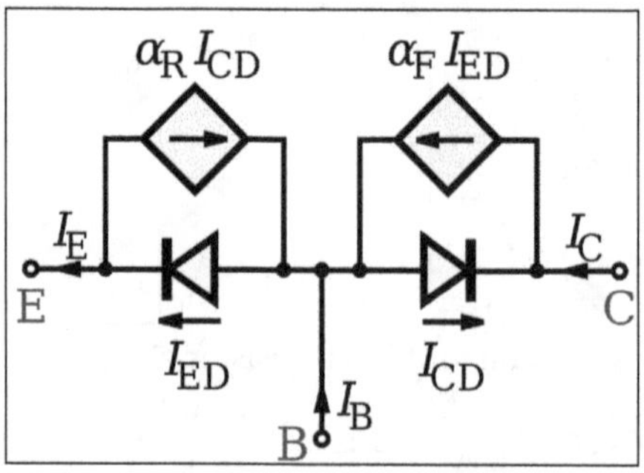

Ebers–Moll model for an NPN transistor. IB, IC and IE are the base, collector and emitter currents; ICD and IED are the collector and emitter diode currents; αF and αR are the forward and reverse common-base current gains.

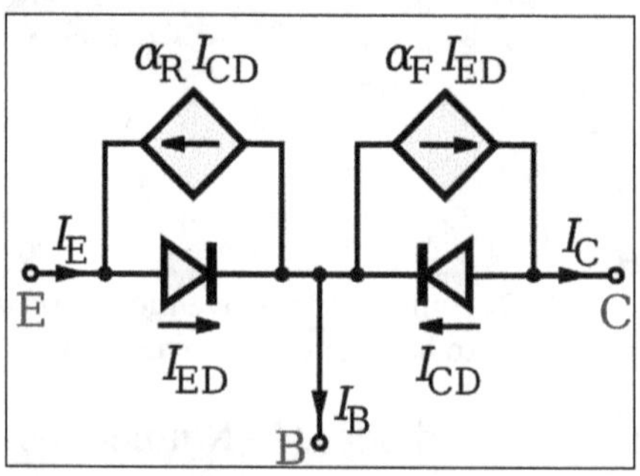

Ebers–Moll model for a PNP transistor.

The DC emitter and collector currents in active mode are well modeled by an approximation to the Ebers–Moll model:

$$I_E = I_{ES}\left(e^{\frac{V_{BE}}{V_T}} - 1\right)$$

$$I_C = \alpha_F I_E$$

$$I_B = (1-\alpha_F) I_E$$

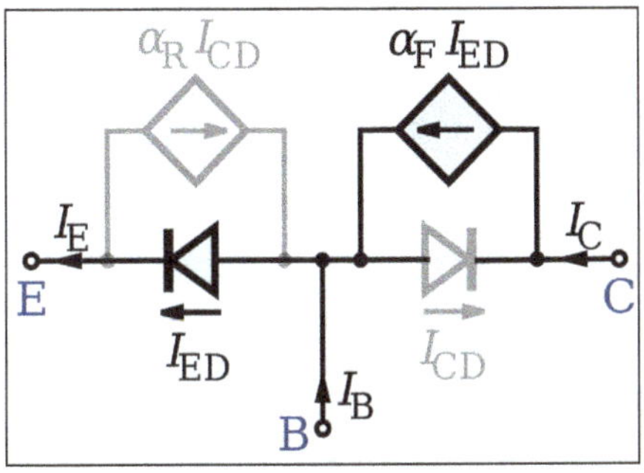

Approximated Ebers–Moll model for an NPN transistor in the forward active mode. The collector diode is reverse-biased so ICD is virtually zero. Most of the emitter diode current (αF is nearly 1) is drawn from the collector, providing the amplification of the base current.

The base internal current is mainly by diffusion and

$$J_{n(base)} = \frac{1}{W} qD_n n_{bo} e^{\frac{V_{EB}}{V_T}}$$

where,

- V_T is the thermal voltage kT/q (approximately 26 mV at 300 K ≈ room temperature).
- I_E is the emitter current.
- I_C is the collector current.
- α_F is the common base forward short-circuit current gain (0.98 to 0.998).
- is the reverse saturation current of the base–emitter diode (on the order of 10^{-15} to 10^{-12} amperes).
- I_{ES} is the base–emitter voltage.
- V_{BE} is the diffusion constant for electrons in the p-type base.
- *W* is the base width.

The α and forward β parameters are as described previously. A reverse β is sometimes included in the model.

The unapproximated Ebers–Moll equations used to describe the three currents in any operating region are given below. These equations are based on the transport model for a bipolar junction transistor.

$$i_C = I_S \left[\left(e^{\frac{V_{BE}}{V_T}} - e^{\frac{V_{BC}}{V_T}} \right) - \frac{1}{\beta_R} \left(e^{\frac{V_{BC}}{V_T}} - 1 \right) \right]$$

$$i_B = I_S\left[\frac{1}{\beta_F}\left(e^{\frac{V_{BE}}{V_T}} - 1\right) + \frac{1}{\beta_R}\left(e^{\frac{V_{BC}}{V_T}} - 1\right)\right]$$

$$i_E = I_S\left[\left(e^{\frac{V_{BE}}{V_T}} - e^{\frac{V_{BC}}{V_T}}\right) + \frac{1}{\beta_F}\left(e^{\frac{V_{BE}}{V_T}} - 1\right)\right]$$

where

- i_C is the collector current.
- i_B is the base current.
- i_E is the emitter current.
- β_F is the forward common emitter current gain (20 to 500),
- β_R is the reverse common emitter current gain (0 to 20),
- I_S is the reverse saturation current (on the order of 10^{-15} to 10^{-12} amperes),
- V_T is the thermal voltage (approximately 26 mV at 300 K $\approx$ room temperature),
- V_{BE} is the base–emitter voltage,
- V_{BC} is the base–collector voltage.

Base-width Modulation

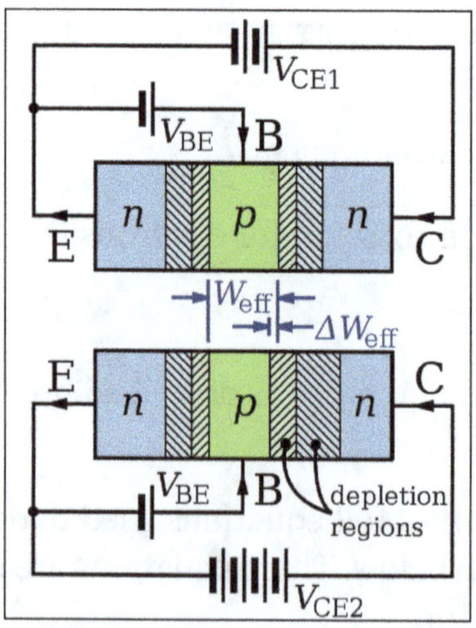

Top: NPN base width for low collector–base reverse bias; Bottom: narrower NPN base width for large collector–base reverse bias. Hashed regions are depleted regions.

As the collector–base voltage ($V_{CB} = V_{CE} - V_{BE}$) varies, the collector–base depletion region varies in size. An increase in the collector–base voltage, for example, causes a greater reverse bias across the collector–base junction, increasing the collector–base depletion region width, and decreasing the width of the base. This variation in base width often is called the *Early effect* after its discoverer James M. Early.

Narrowing of the base width has two consequences:

- There is a lesser chance for recombination within the "smaller" base region.
- The charge gradient is increased across the base, and consequently, the current of minority carriers injected across the emitter junction increases.

Both factors increase the collector or "output" current of the transistor in response to an increase in the collector–base voltage.

In the forward-active region, the Early effect modifies the collector current (i_C) and the forward common emitter current gain (β_F) as given by:

$$i_C = I_S e^{\frac{V_{BE}}{V_T}} \left(1 + \frac{V_{CE}}{V_A}\right)$$

$$\beta_F = \beta_{F0} \left(1 + \frac{V_{CB}}{V_A}\right)$$

$$r_o = \frac{V_A}{I_C}$$

where:

- V_{CE} is the collector–emitter voltage.
- V_A is the Early voltage (15 V to 150 V).
- β_{F0} is forward common-emitter current gain when $V_{CB} = 0$ V.
- r_o is the output impedance.
- I_C is the collector current.

Punchthrough

When the base–collector voltage reaches a certain (device-specific) value, the base–collector depletion region boundary meets the base–emitter depletion region boundary. When in this state the transistor effectively has no base. The device thus loses all gain when in this state.

Gummel–Poon Charge-control Model

The Gummel–Poon model is a detailed charge-controlled model of BJT dynamics, which has been adopted and elaborated by others to explain transistor dynamics in greater detail than the terminal-based models typically do. This model also includes the dependence of transistor β-values upon the direct current levels in the transistor, which are assumed current-independent in the Ebers–Moll model.

Small-signal Models

Hybrid-pi Model

The hybrid-pi model is a popular circuit model used for analyzing the small signal and AC behavior of bipolar junction and field effect transistors. Sometimes it is also called Giacoletto model because it was introduced by L.J. Giacoletto in 1969. The model can be quite accurate for low-frequency circuits and can easily be adapted for higher-frequency circuits with the addition of appropriate inter-electrode capacitances and other parasitic elements.

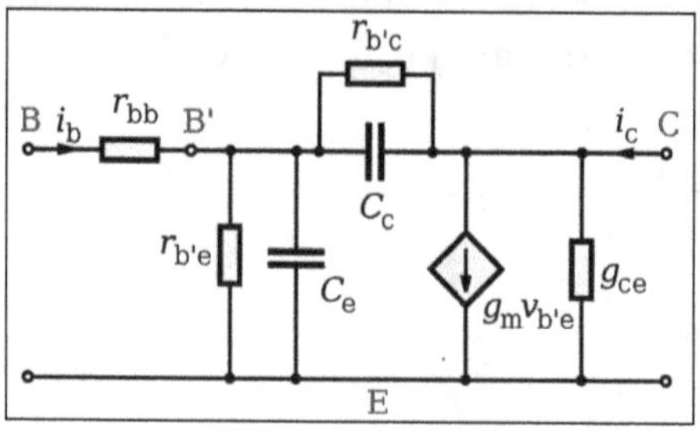

Hybrid-pi model.

h-parameter Model

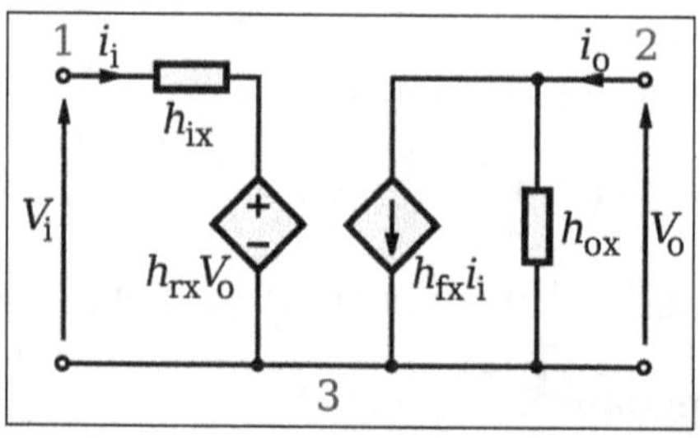

Generalized h-parameter model of an NPN BJT. Replace x with e, b or c for CE, CB and CC topologies respectively.

Another model commonly used to analyze BJT circuits is the h-parameter model, closely related to the hybrid-pi model and the y-parameter two-port, but using input current and output voltage as independent variables, rather than input and output voltages. This two-port network is particularly suited to BJTs as it lends itself easily to the analysis of circuit behaviour, and may be used to develop further accurate models. As shown, the term, x, in the model represents a different BJT lead depending on the topology used. For common-emitter mode the various symbols take on the specific values as:

- Terminal 1, base.
- Terminal 2, collector.

- Terminal 3 (common), emitter; giving x to be e.
- i_i, base current (i_b).
- i_o, collector current (i_c).
- V_{in}, base-to-emitter voltage (V_{BE}).
- V_o, collector-to-emitter voltage (V_{CE}).

The h-parameters are given by:

- $h_{ix} = h_{ie}$, the input impedance of the transistor (corresponding to the base resistance r_{pi}).
- $h_{rx} = h_{re}$, represents the dependence of the transistor's I_B–V_{BE} curve on the value of V_{CE}. It is usually very small and is often neglected (assumed to be zero).
- $h_{fx} = h_{fe}$, the current-gain of the transistor. This parameter is often specified as h_{FE} or the DC current-gain (β_{DC}) in datasheets.
- $h_{ox} = 1/h_{oe}$, the output impedance of transistor. The parameter h_{oe} usually corresponds to the output admittance of the bipolar transistor and has to be inverted to convert it to an impedance.

The h-parameters have lower-case subscripts and hence signify AC conditions or analyses. For DC conditions they are specified in upper-case. For the CE topology, an approximate h-parameter model is commonly used which further simplifies the circuit analysis. For this the h_{oe} and h_{re} parameters are neglected (that is, they are set to infinity and zero, respectively). The h-parameter model as shown is suited to low-frequency, small-signal analysis. For high-frequency analyses the inter-electrode capacitances that are important at high frequencies must be added.

Etymology of h_{FE}

The h refers to its being an h-parameter, a set of parameters named for their origin in a hybrid equivalent circuit model. F is from forward current amplification also called the current gain. E refers to the transistor operating in a common emitter (CE) configuration. Capital letters used in the subscript indicate that h_{FE} refers to a direct current circuit.

Industry Models

The Gummel–Poon SPICE model is often used, but it suffers from several limitations. These have been addressed in various more advanced models: Mextram, VBIC, HICUM, Modella.

Applications

The BJT remains a device that excels in some applications, such as discrete circuit design, due to the very wide selection of BJT types available, and because of its high transconductance and output resistance compared to MOSFETs.

The BJT is also the choice for demanding analog circuits, especially for very-high-frequency applications, such as radio-frequency circuits for wireless systems.

- High-speed digital logic:
    - Emitter-coupled logic (ECL) use BJTs.
    - Bipolar transistors can be combined with MOSFETs in an integrated circuit by using a BiCMOS process of wafer fabrication to create circuits that take advantage of the application strengths of both types of transistor.
- Amplifiers: The transistor parameters α and β characterizes the current gain of the BJT. It is this gain that allows BJTs to be used as the building blocks of electronic amplifiers. The three main BJT amplifier topologies are:
    - Common emitter.
    - Common base.
    - Common collector.
- Temperature sensors: Because of the known temperature and current dependence of the forward-biased base–emitter junction voltage, the BJT can be used to measure temperature by subtracting two voltages at two different bias currents in a known ratio.
- Logarithmic converters: Because base–emitter voltage varies as the logarithm of the base–emitter and collector–emitter currents, a BJT can also be used to compute logarithms and anti-logarithms. A diode can also perform these nonlinear functions but the transistor provides more circuit flexibility.

Vulnerabilities

Exposure of the transistor to ionizing radiation causes radiation damage. Radiation causes a buildup of 'defects' in the base region that act as recombination centers. The resulting reduction in minority carrier lifetime causes gradual loss of gain of the transistor.

Transistors have "maximum ratings", including power ratings (essentially limited by self-heating), maximum collector and base currents (both continuous/DC ratings and peak), and breakdown voltage ratings, beyond which the device may fail or at least perform badly.

In addition to normal breakdown ratings of the device, power BJTs are subject to a failure mode called secondary breakdown, in which excessive current and normal imperfections in the silicon die cause portions of the silicon inside the device to become disproportionately hotter than the others. The electrical resistivity of doped silicon, like other semiconductors, has a negative temperature coefficient, meaning that it conducts more current at higher temperatures. Thus, the hottest part of the die conducts the most current, causing its conductivity to increase, which then causes it to become progressively hotter again, until the device fails internally. The thermal runaway process associated with secondary breakdown, once triggered, occurs almost instantly and may catastrophically damage the transistor package.

If the emitter-base junction is reverse biased into avalanche or Zener mode and charge flows for a short period of time, the current gain of the BJT will be permanently degraded.

Field-effect Transistor

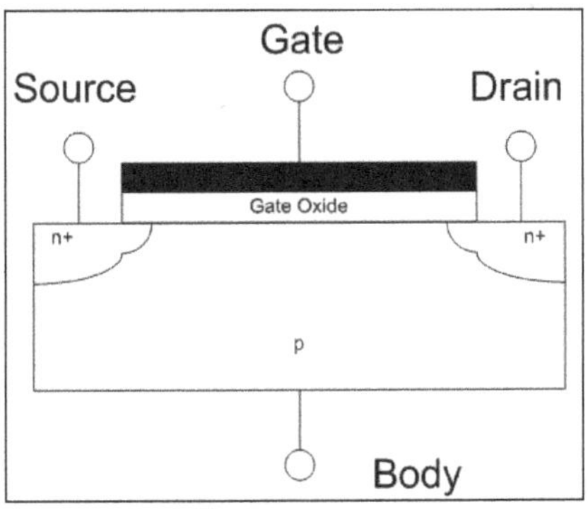

Cross-sectional view of a field-effect transistor, showing source, gate and drain terminals.

The field-effect transistor (FET) is a type of transistor which uses an electric field to control the flow of current. FETs are devices with three terminals: source, gate, and drain. FETs control the flow of current by the application of a voltage to the gate, which in turn alters the conductivity between the drain and source.

FETs are also known as unipolar transistors since they involve single-carrier-type operation. That is, FETs use electrons or holes as charge carriers in their operation, but not both. Many different types of field effect transistors exist. Field effect transistors generally display very high input impedance at low frequencies. The most widely used field-effect transistor is the MOSFET (metal-oxide-semiconductor field-effect transistor).

FETs can be majority-charge-carrier devices, in which the current is carried predominantly by majority carriers, or minority-charge-carrier devices, in which the current is mainly due to a flow of minority carriers. The device consists of an active channel through which charge carriers, electrons or holes, flow from the source to the drain. Source and drain terminal conductors are connected to the semiconductor through ohmic contacts. The conductivity of the channel is a function of the potential applied across the gate and source terminals.

The FET's three terminals are:

- Source (S), through which the carriers enter the channel. Conventionally, current entering the channel at S is designated by I_S.
- Drain (D), through which the carriers leave the channel. Conventionally, current entering the channel at D is designated by I_D. Drain-to-source voltage is V_{DS}.
- Gate (G), the terminal that modulates the channel conductivity. By applying voltage to G, one can control I_D.

More about Terminals

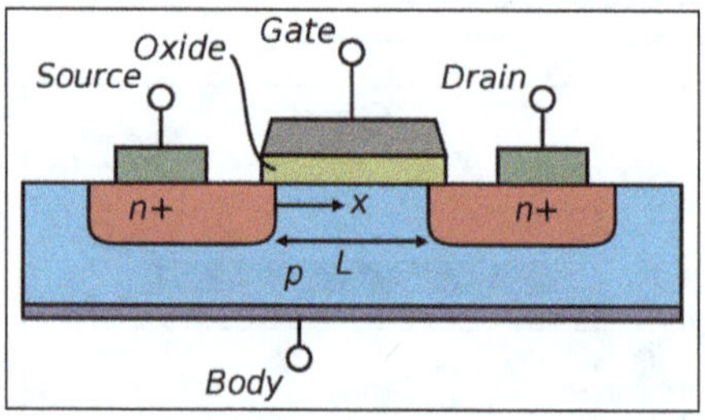

Cross section of an n-type MOSFET.

All FETs have source, drain, and gate terminals that correspond roughly to the emitter, collector, and base of BJTs. Most FETs have a fourth terminal called the body, base, bulk, or substrate. This fourth terminal serves to bias the transistor into operation; it is rare to make non-trivial use of the body terminal in circuit designs, but its presence is important when setting up the physical layout of an integrated circuit. The size of the gate, length L, is the distance between source and drain. The width is the extension of the transistor, in the direction perpendicular to the cross section (i.e., into/out of the screen). Typically the width is much larger than the length of the gate. A gate length of 1 µm limits the upper frequency to about 5 GHz, 0.2 µm to about 30 GHz.

The names of the terminals refer to their functions. The gate terminal may be thought of as controlling the opening and closing of a physical gate. This gate permits electrons to flow through or blocks their passage by creating or eliminating a channel between the source and drain. Electron-flow from the source terminal towards the drain terminal is influenced by an applied voltage. The body simply refers to the bulk of the semiconductor in which the gate, source and drain lie. Usually the body terminal is connected to the highest or lowest voltage within the circuit, depending on the type of the FET. The body terminal and the source terminal are sometimes connected together since the source is often connected to the highest or lowest voltage within the circuit, although there are several uses of FETs which do not have such a configuration, such as transmission gates and cascode circuits.

Effect of Gate Voltage on Current

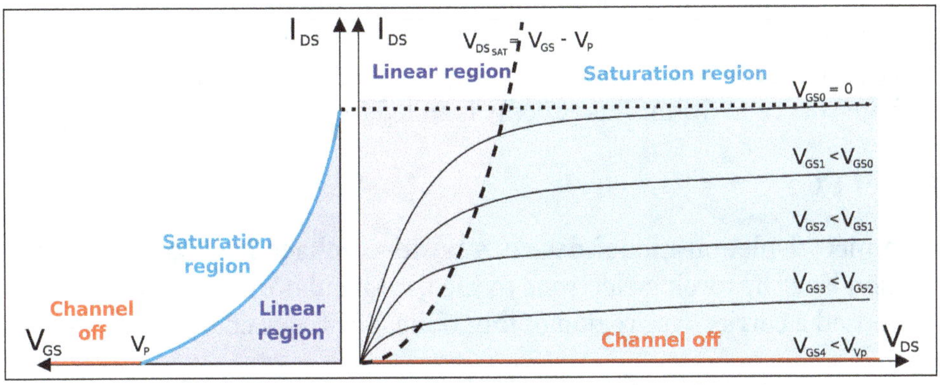

I–V characteristics and output plot of a JFET n-channel transistor.

The FET controls the flow of electrons (or electron holes) from the source to drain by affecting the size and shape of a "conductive channel" created and influenced by voltage (or lack of voltage) applied across the gate and source terminals. This conductive channel is the "stream" through which electrons flow from source to drain.

n-channel FET

In an n-channel "depletion-mode" device, a negative gate-to-source voltage causes a depletion region to expand in width and encroach on the channel from the sides, narrowing the channel. If the active region expands to completely close the channel, the resistance of the channel from source to drain becomes large, and the FET is effectively turned off like a switch. This is called "pinch-off", and the voltage at which it occurs is called the "pinch-off voltage". Conversely, a positive gate-to-source voltage increases the channel size and allows electrons to flow easily.

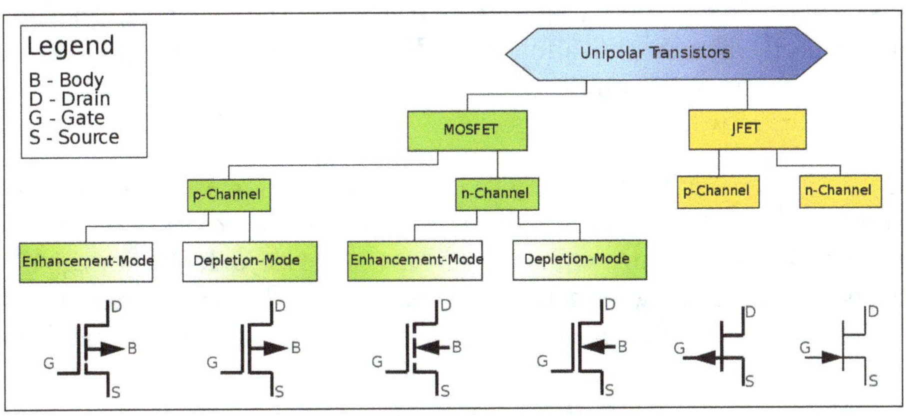

FET conventional symbol types.

In an n-channel "enhancement-mode" device, a conductive channel does not exist naturally within the transistor, and a positive gate-to-source voltage is necessary to create one. The positive voltage attracts free-floating electrons within the body towards the gate, forming a conductive channel. But first, enough electrons must be attracted near the gate to counter the dopant ions added to the body of the FET; this forms a region with no mobile carriers called a depletion region, and the voltage at which this occurs is referred to as the threshold voltage of the FET. Further gate-to-source voltage increase will attract even more electrons towards the gate which are able to create a conductive channel from source to drain; this process is called inversion.

p-channel FET

In a p-channel "depletion-mode" device, a positive voltage from gate to body widens the depletion layer by forcing electrons to the gate-insulator/semiconductor interface, leaving exposed a carrier-free region of immobile, positively charged acceptor ions.

Conversely, in a p-channel "enhancement-mode" device, a conductive region does not exist and negative voltage must be used to generate a conduction channel.

Effect of Drain-to-source Voltage on Channel

For either enhancement- or depletion-mode devices, at drain-to-source voltages much less than gate-to-source voltages, changing the gate voltage will alter the channel resistance, and drain current will be proportional to drain voltage. In this mode the FET operates like a variable resistor and the FET is said to be operating in a linear mode or ohmic mode.

If drain-to-source voltage is increased, this creates a significant asymmetrical change in the shape of the channel due to a gradient of voltage potential from source to drain. The shape of the inversion region becomes "pinched-off" near the drain end of the channel. If drain-to-source voltage is increased further, the pinch-off point of the channel begins to move away from the drain towards the source. The FET is said to be in saturation mode; although some authors refer to it as active mode, for a better analogy with bipolar transistor operating regions. The saturation mode, or the region between ohmic and saturation, is used when amplification is needed. The in-between region is sometimes considered to be part of the ohmic or linear region, even where drain current is not approximately linear with drain voltage.

Even though the conductive channel formed by gate-to-source voltage no longer connects source to drain during saturation mode, carriers are not blocked from flowing. Considering again an n-channel enhancement-mode device, a depletion region exists in the p-type body, surrounding the conductive channel and drain and source regions. The electrons which comprise the channel are free to move out of the channel through the depletion region if attracted to the drain by drain-to-source voltage. The depletion region is free of carriers and has a resistance similar to silicon. Any increase of the drain-to-source voltage will increase the distance from drain to the pinch-off point, increasing the resistance of the depletion region in proportion to the drain-to-source voltage applied. This proportional change causes the drain-to-source current to remain relatively fixed, independent of changes to the drain-to-source voltage, quite unlike its ohmic behavior in the linear mode of operation. Thus, in saturation mode, the FET behaves as a constant-current source rather than as a resistor, and can effectively be used as a voltage amplifier. In this case, the gate-to-source voltage determines the level of constant current through the channel.

Composition

FETs can be constructed from various semiconductors—silicon is by far the most common. Most FETs are made by using conventional bulk semiconductor processing techniques, using a single crystal semiconductor wafer as the active region, or channel.

Among the more unusual body materials are amorphous silicon, polycrystalline silicon or other amorphous semiconductors in thin-film transistors or organic field-effect transistors (OFETs) that are based on organic semiconductors; often, OFET gate insulators and electrodes are made of organic materials, as well. Such FETs are manufactured using a variety of materials such as silicon carbide (SiC), gallium arsenide (GaAs), gallium nitride (GaN), and indium gallium arsenide (InGaAs). In June 2011, IBM announced that it had successfully used graphene-based FETs in an integrated circuit. These transistors are capable of about 2.23 GHz cutoff frequency, much higher than standard silicon FETs.

Types

The channel of a FET is doped to produce either an n-type semiconductor or a p-type semiconductor. The drain and source may be doped of opposite type to the channel, in the case of enhancement mode FETs, or doped of similar type to the channel as in depletion mode FETs. Field-effect transistors are also distinguished by the method of insulation between channel and gate. Types of FETs include:

- The MOSFET (metal–oxide–semiconductor field-effect transistor) utilizes an insulator (typically SiO_2) between the gate and the body. This is by far the most common type of FET.
    - The DGMOSFET (dual-gate MOSFET) or DGMOS, a MOSFET with two insulated gates.
    - The IGBT (insulated-gate bipolar transistor) is a device for power control. It has a structure akin to a MOSFET coupled with a bipolar-like main conduction channel. These are commonly used for the 200–3000 V drain-to-source voltage range of operation. Power MOSFETs are still the device of choice for drain-to-source voltages of 1 to 200 V.

  - The MNOS (metal–nitride–oxide–semiconductor transistor) utilizes an nitride-oxide layer insulator between the gate and the body.
  - The ISFET (ion-sensitive field-effect transistor) can be used to measure ion concentrations in a solution; when the ion concentration (such as H^+) changes, the current through the transistor will change accordingly.
  - The BioFET (Biologically sensitive field-effect transistor) is a class of sensors/biosensors based on ISFET technology which are utilized to detect charged molecules; when a charged molecule is present, changes in the electrostatic field at the BioFET surface result in a measurable change in current through the transistor. These include enzyme modified FETs (EnFETs), immunologically modified FETs (ImmunoFETs), gene-modified FETs (GenFETs), DNAFETs, cell-based BioFETs (CPFETs), beetle/chip FETs (BeetleFETs), and FETs based on ion-channels/protein binding.
  - The DNAFET (DNA field-effect transistor) is a specialized FET that acts as a biosensor, by using a gate made of single-strand DNA molecules to detect matching DNA strands.
- The JFET (junction field-effect transistor) uses a reverse biased p–n junction to separate the gate from the body.
  - The static induction transistor (SIT) is a type of JFET with a short channel.
- The DEPFET is a FET formed in a fully depleted substrate and acts as a sensor, amplifier and memory node at the same time. It can be used as an image (photon) sensor.
- The FREDFET (fast-reverse or fast-recovery epitaxial diode FET) is a specialized FET designed to provide a very fast recovery (turn-off) of the body diode, making it convenient for driving inductive loads such as electric motors, especially medium-powered brushless DC motors.
- The HIGFET (heterostructure insulated-gate field-effect transistor) is now used mainly in research.
- The MODFET (modulation-doped field-effect transistor) is a high-electron-mobility transistor using a quantum well structure formed by graded doping of the active region.
- The TFET (tunnel field-effect transistor) is based on band-to-band tunneling.
- The HEMT (high-electron-mobility transistor), also called a HFET (heterostructure FET), can be made using bandgap engineering in a ternary semiconductor such as AlGaAs. The fully depleted wide-band-gap material forms the isolation between gate and body.

- The MESFET (metal–semiconductor field-effect transistor) substitutes the p–n junction of the JFET with a Schottky barrier; and is used in GaAs and other III-V semiconductor materials.
- The NOMFET is a nanoparticle organic memory field-effect transistor.
- The GNRFET (graphene nanoribbon field-effect transistor) uses a graphene nanoribbon for its channel.
- The VeSFET (vertical-slit field-effect transistor) is a square-shaped junctionless FET with a narrow slit connecting the source and drain at opposite corners. Two gates occupy the other corners, and control the current through the slit.
- The CNTFET (carbon nanotube field-effect transistor).
- The OFET (organic field-effect transistor) uses an organic semiconductor in its channel.
- The QFET (quantum field effect transistor) takes advantage of quantum tunneling to greatly increase the speed of transistor operation by eliminating the traditional transistor's area of electron conduction.
- The SB-FET (Schottky-barrier field-effect transistor) is a field-effect transistor with metallic source and drain contact electrodes, which create Schottky barriers at both the source-channel and drain-channel interfaces.
- The GFET is a highly sensitive graphene-based field effect transistor used as biosensors and chemical sensors. Due to the 2 dimensional structure of graphene, along with its physical properties, GFETs offer increased sensitivity, and reduced instances of 'false positives' in sensing applications
- The Fe FET uses a ferroelectric between the gate, allowing the transistor to retain its state in the absence of bias - such devices may have application as non-volatile memory.

Advantages

The FET has high gate-to-main current resistance, on the order of 100 MΩ or more, providing a high degree of isolation between control and flow. Because base current noise will increase with shaping time, a FET typically produces less noise than a bipolar junction transistor (BJT), and is found in noise-sensitive electronics such as tuners and low-noise amplifiers for VHF and satellite receivers. It is relatively immune to radiation. It exhibits no offset voltage at zero drain current and makes an excellent signal chopper. It typically has better thermal stability than a BJT. Because they are controlled by gate charge, once the gate is closed or open, there is no additional power draw, as there would be with a bipolar junction transistor or with non-latching relays in some states. This allows extremely low-power switching, which in turn allows greater miniaturization of circuits because heat dissipation needs are reduced compared to other types of switches.

Disadvantages

A field-effect transistor has a relatively low gain–bandwidth product compared to a BJT. The MOSFET is very susceptible to overload voltages, thus requiring special handling during installation. The fragile insulating layer of the MOSFET between the gate and channel makes it vulnerable to electrostatic discharge or changes to threshold voltage during handling. This is not usually a problem after the device has been installed in a properly designed circuit.

FETs often have a very low "on" resistance and have a high "off" resistance. However, the intermediate resistances are significant, and so FETs can dissipate large amounts of power while switching. Thus efficiency can put a premium on switching quickly, but this can cause transients that can excite stray inductances and generate significant voltages that can couple to the gate and cause unintentional switching. FET circuits can therefore require very careful layout and can involve trades between switching speed and power dissipation. There is also a trade-off between voltage rating and "on" resistance, so high-voltage FETs have a relatively high "on" resistance and hence conduction losses.

Failure Modes

FETs are relatively robust, especially when operated within the temperature and electrical limitations defined by the manufacturer (proper derating). However, modern FET devices can often incorporate a body diode. If the characteristics of the body diode are not taken into consideration, the FET can experience slow body diode behavior, where a parasitic transistor will turn on and allow high current to be drawn from drain to source when the FET is off.

Uses

The most commonly used FET is the MOSFET. The CMOS (complementary metal oxide semiconductor) process technology is the basis for modern digital integrated circuits. This process technology uses an arrangement where the (usually "enhancement-mode") p-channel MOSFET and n-channel MOSFET are connected in series such that when one is on, the other is off.

In FETs, electrons can flow in either direction through the channel when operated in the linear mode. The naming convention of drain terminal and source terminal is somewhat arbitrary, as the devices are typically (but not always) built symmetrical from source to drain. This makes FETs suitable for switching analog signals between paths (multiplexing). With this concept, one can construct a solid-state mixing board, for example. FET is commonly used as an amplifier. For example, due to its large input resistance and low output resistance, it is effective as a buffer in common-drain (source follower) configuration.

IGBTs are used in switching internal combustion engine ignition coils, where fast switching and voltage blocking capabilities are important.

Source-gated Transistor

Source-gated transistors are more robust to manufacturing and environmental issues in large-area electronics such as display screens, but are slower in operation than FETs.

References

- An-introduction-to-analog-electronics: engineersgarage.com, Retrieved 13 April, 2020
- Analog-circuits, topics-engineering: sciencedirect.com, Retrieved 02 August, 2020
- What-is-jfet-basics-construction-working-and-biasing: circuitdigest.com, Retrieved 19 March, 2020
- Agarwal, Anant; Lang, Jeffrey (2005). Foundations of Analog and Digital Electronic Circuits. Morgan Kaufmann. p. 331. ISBN 978-0080506814
- Peter Ashburn (2003). SiGe Heterojunction Bipolar Transistors. New York: Wiley. Chapter 10. ISBN 978-0-470-84838-8
- Lojek, Bo (2007). History of Semiconductor Engineering. Springer Science & Business Media. p. 120. ISBN 9783540342588
- Bakshi, U. A.; Godse, A. P. (2007). "§8.2 The depletion mode MOSFET". Electronic Circuits. Technical Publications. p. 812. ISBN 978-81-8431-284-3

4
Transducers

Transducers are the devices that convert a signal in one form of energy to another. A few of its types are active transducer, passive transducer, displacement transducer, pressure transducer, flow transducer, etc. All the types and concepts of transducers have been carefully analyzed in this chapter.

A transducer is a device that converts energy from one form to another. Usually a transducer converts a signal in one form of energy to a signal in another.

Transducers are often employed at the boundaries of automation, measurement, and control systems, where electrical signals are converted to and from other physical quantities (energy, force, torque, light, motion, position, etc). The process of converting one form of energy to another is known as transduction.

Active Transducers

Tachogenerator

A tachogenerator, or tachometer generator, is an electromechanical device used to accurately measure the speed of engines and motors. In fact, a tachogenerator can measure the speed and direction of all manner of rotational devices once connected and will output a voltage proportional to the rotation around its own shaft. In other words, a tachogenerator can convert mechanical energy into electrical energy. A tachogenerator is commonly used to handle voltages between 0 and 10 volts and is classed as a precision instrument. We supply solid shaft, blind shaft and hollow shaft types of tachogenerator.

- Solid Shaft Tachogenerator: Solid shaft tachogenerators are connected to a shaft which, in turn, is connected to an external rotational device. It is the rotation of the external device which turns the tachogenerator shaft and thus generating a very specific range of voltages in accordance with the speed and direction of the shaft. Tachogenerators can indicate rotational direction due to the fact that if a tachogenerator shaft is reversed, the output voltage polarity will change. The solid-shaft tachogenerator is best suited to high load applications.

- Hollow Shaft: A Hollow-shaft tachogenerator differs internally from solid shaft tachogenerators in that they are designed with four magnetic poles rather than two. This design allows for the tachogenerator to work with lower voltage loads. An example of the use of a DC hollow shaft tachogenerator is to sense the speed of an elevator. The DC tachogenerator is installed in the hoisting equipment on the traction sheave that drives the cables. It enables the accurate control of cable speed to make sure that the lift stops at the right floor and does so smoothly.
- Tachogenerator Applications: A tachogenerator can be found in various kinds of machine tool and other equipment where the measurement of speed and direction is essential. These include electric motors, conveyor belts, engines, fans and mixers. Consumer products where you would expect to find a tachogenerator would be in a washing machine or a food processor. A tachogenerator is also used in voltage control and current limiting circuits to protect equipment from overheating or to protect other components within an electromechanical device.

Thermocouple

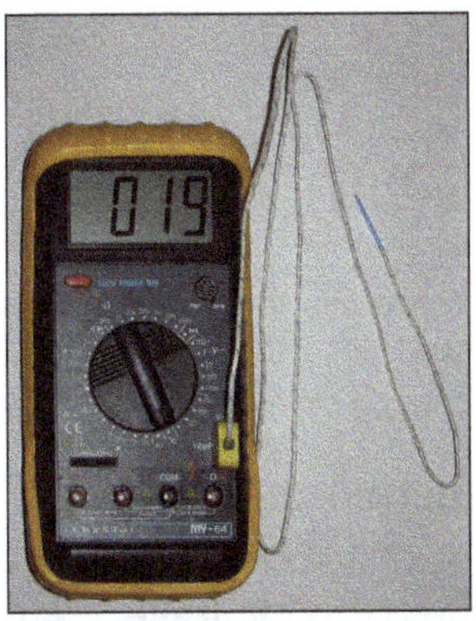

Thermocouple connected to a multimeter displaying room temperature in °C.

A thermocouple is an electrical device consisting of two dissimilar electrical conductors forming an electrical junction. A thermocouple produces a temperature-dependent voltage as a result of the thermoelectric effect, and this voltage can be interpreted to measure temperature. Thermocouples are a widely used type of temperature sensor.

Commercial thermocouples are inexpensive, interchangeable, are supplied with standard connectors, and can measure a wide range of temperatures. In contrast to most other methods of temperature measurement, thermocouples are self powered and require no external form of excitation. The main limitation with thermocouples is precision; system errors of less than one degree Celsius (°C) can be difficult to achieve.

Thermocouples are widely used in science and industry. Applications include temperature measurement for kilns, gas turbine exhaust, diesel engines, and other industrial processes. Thermocouples are also used in homes, offices and businesses as the temperature sensors in thermostats, and also as flame sensors in safety devices for gas-powered appliances.

Principle of Operation

In 1821, the German physicist Thomas Johann Seebeck discovered that when different metals are joined at the ends and there is a temperature difference between the joints, a magnetic field is observed. At the time, Seebeck referred to this consequence as thermo-magnetism. The magnetic field he observed was later shown to be due to thermo-electric current. In practical use, the voltage generated at a single junction of two different types of wire is what is of interest as this can be used to measure temperature at very high and low temperatures. The magnitude of the voltage depends on the types of wire being used. Generally, the voltage is in the microvolt range and care must be taken to obtain a usable measurement. Although very little current flows, power can be generated by a single thermocouple junction. Power generation using multiple thermocouples, as in a thermopile, is common.

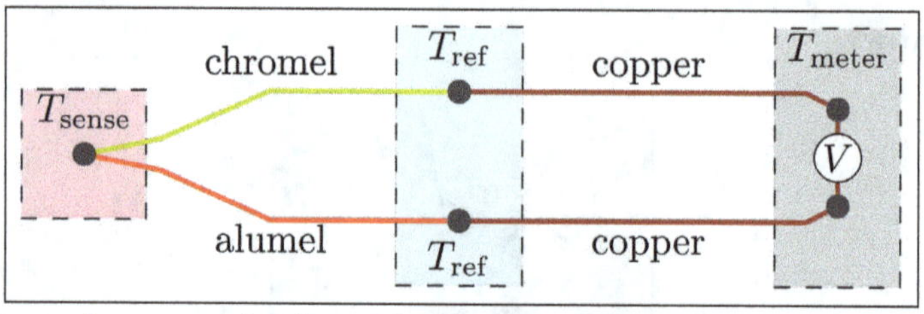

K-type thermocouple (chromel–alumel) in the standard thermocouple measurement configuration. The measured voltage V can be used to calculate temperature $T_{sense,}$ provided that temperature T_{ref} is known.

The standard configuration for thermocouple usage is shown in the figure. Briefly, the desired temperature T_{sense} is obtained using three inputs—the characteristic function $E(T)$ of the thermocouple, the measured voltage V, and the reference junctions' temperature T_{ref}. The solution to the equation $E(T_{sense}) = V + E(T_{ref})$ yields T_{sense}. These details are often hidden from the user since the reference junction block (with T_{ref} thermometer), voltmeter, and equation solver are combined into a single product.

Physical Principle: Seebeck Effect

The Seebeck effect refers to an electromotive force whenever there is a temperature gradient in a conductive material. Under open-circuit conditions where there is no internal current flow, the gradient of voltage (∇V) is directly proportional to the gradient in temperature (∇T):

$$\nabla V = -S(T)\nabla T,$$

where $S(T)$ is a temperature-dependent material property known as the Seebeck coefficient.

The standard measurement configuration shown in the figure shows four temperature regions and thus four voltage contributions:

- Change from T_{meter} to T_{ref}, in the lower copper wire.
- Change from T_{ref} to T_{sense}, in the alumel wire.
- Change from T_{sense} to T_{ref}, in the chromel wire.
- Change from T_{ref} to T_{meter}, in the upper copper wire.

The first and fourth contributions cancel out exactly, because these regions involve the same temperature change and an identical material. As a result, T_{meter} does not influence the measured voltage. The second and third contributions do not cancel, as they involve different materials.

The measured voltage turns out to be,

$$V = \int_{T_{ref}}^{T_{sense}} (S_+(T) - S_-(T))dT,$$

where S_+ and S_- are the Seebeck coefficients of the conductors attached to the positive and negative terminals of the voltmeter, respectively (chromel and alumel in the figure).

Characteristic Function

An integral does not need to be performed for every temperature measurement. Rather, the thermocouple's behaviour is captured by a characteristic function $E(T)$, which needs only to be consulted at two arguments,

$$V = E(T_{sense}) - E(T_{ref}).$$

In terms of the Seebeck coefficients, the characteristic function is defined by,

$$E(T) = \int^{T} S_+(T') - S_-(T')dT' + \text{const}$$

The constant of integration in this indefinite integral has no significance, but is conventionally chosen such that $E(0\,^{\circ}C) = 0$.

Thermocouple manufacturers and metrology standards organizations such as NIST provide tables of the function $E(T)$ that have been measured and interpolated over a range of temperatures, for particular thermocouple types.

Requirement for a Reference Junction

Reference junction block inside a Fluke CNX t3000 temperature meter. Two white wires connect to a thermistor (embedded in white thermal compound) to measure the reference junctions' temperature.

To obtain the desired measurement of T_{sense}, it is not sufficient to just measure V. The temperature at the reference junctions T_{ref} must be already known. Two strategies are often used here:

- "Ice bath" method: The reference junction block is immersed in a semi-frozen bath of distilled water at atmospheric pressure. The precise temperature of the melting point phase transition acts as a natural thermostat, fixing T_{ref} to 0 °C.
- Reference junction sensor (known as "cold junction compensation"): The reference junction block is allowed to vary in temperature, but the temperature is measured at this block using a separate temperature sensor. This secondary measurement is used to compensate for temperature variation at the junction block. The thermocouple junction is often exposed to extreme environments, while the reference junction is often mounted near the instrument's location. Semiconductor thermometer devices are often used in modern thermocouple instruments.

In both cases the value $V + E(T_{ref})$ is calculated, then the function $E(T)$ is searched for a matching value. The argument where this match occurs is the value of T_{sense}.

Practical Concerns

Thermocouples ideally should be very simple measurement devices, with each type being characterized by a precise $E(T)$ curve, independent of any other details. In reality, thermocouples are affected by issues such as alloy manufacturing uncertainties, aging effects, and circuit design mistakes/misunderstandings.

Circuit Construction

A common error in thermocouple construction is related to cold junction compensation. If an error is made on the estimation of T_{ref}, the same error will be carried over to the temperature measurement. For the simplest measurements, thermocouple wires are connected to copper far away from the hot or cold point whose temperature is measured; the cold junction is then assumed to be at room temperature, but that temperature can vary.

Junctions should be made in a reliable manner, but there are many possible approaches to accomplish this. For low temperatures, junctions can be brazed or soldered; however, it may be difficult to find a suitable flux and this may not be suitable at the sensing junction due to the solder's low melting point. Reference and extension junctions are therefore usually made with screw terminal blocks. For high temperatures, a common approach is a spot weld or crimp using a durable material.

A common myth regarding thermocouples is that junctions must be made cleanly without involving a third metal, to avoid unwanted added EMFs. This may result from another common misunderstanding that the voltage is generated at the junction. In fact, the junctions should in principle have uniform internal temperature; therefore, no voltage is generated at the junction. The voltage is generated in the thermal gradient, along the wire.

A thermocouple produces small signals, often microvolts in magnitude. Precise measurements of this signal require an amplifier with low input offset voltage and with care taken to avoid thermal EMFs from self-heating within the voltmeter itself. If the thermocouple wire has a high resistance for some reason (poor contact at junctions, or very thin wires used for fast thermal response), the measuring instrument should have high input impedance to prevent an offset in the measured voltage. A useful feature in thermocouple instrumentation will simultaneously measure resistance and detect faulty connections in the wiring or at thermocouple junctions.

Metallurgical Grades

While a thermocouple wire type is often described by its chemical composition, the actual aim is to produce a pair of wires that follow a standardized $E(T)$ curve.

Impurities affect each batch of metal differently, producing variable Seebeck coefficients. To match the standard behaviour, thermocouple wire manufacturers will deliberately mix in additional impurities to "dope" the alloy, compensating for uncontrolled variations in source material. As a result, there are standard and specialized grades of thermocouple wire, depending on the level of precision demanded in the thermocouple behaviour. Precision grades may only be available in matched pairs, where one wire is modified to compensate for deficiencies in the other wire.

A special case of thermocouple wire is known as "extension grade", designed to carry the thermoelectric circuit over a longer distance. Extension wires follow the stated $E(T)$ curve but for various reasons they are not designed to be used in extreme environments and so they cannot be used at the sensing junction in some applications. For example, an extension wire may be in a different form, such as highly flexible with stranded construction and plastic insulation, or be part of a multi-wire cable for carrying many thermocouple circuits. With expensive noble metal thermocouples, the extension wires may even be made of a completely different, cheaper material that mimics the standard type over a reduced temperature range.

Aging of Thermocouples

Thermocouples are often used at high temperatures and in reactive furnace atmospheres. In this case, the practical lifetime is limited by thermocouple aging. The thermoelectric coefficients of the wires in a thermocouple that is used to measure very high temperatures may change with time, and the measurement voltage accordingly drops. The simple relationship between the temperature difference of the junctions and the measurement voltage is only correct if each wire is homogeneous (uniform in composition). As thermocouples age in a process, their conductors can lose homogeneity due to chemical and metallurgical changes caused by extreme or prolonged exposure to high temperatures. If the aged section of the thermocouple circuit is exposed to a temperature gradient, the measured voltage will differ, resulting in error.

Aged thermocouples are only partly modified; for example, being unaffected in the parts outside the furnace. For this reason, aged thermocouples cannot be taken out of their installed location and recalibrated in a bath or test furnace to determine error. This also explains why error can sometimes be observed when an aged thermocouple is pulled partly out of a furnace—as the sensor is pulled back, aged sections may see exposure to increased temperature gradients from hot to cold as the aged section now passes through the cooler refractory area, contributing significant error to the measurement. Likewise, an aged thermocouple that is pushed deeper into the furnace might sometimes provide a more accurate reading if being pushed further into the furnace causes the temperature gradient to occur only in a fresh section.

Types

Certain combinations of alloys have become popular as industry standards. Selection of the combination is driven by cost, availability, convenience, melting point, chemical properties, stability, and output. Different types are best suited for different applications. They are usually selected on the basis of the temperature range and sensitivity needed. Thermocouples with low sensitivities (B, R, and S types) have correspondingly lower resolutions. Other selection criteria include the chemical inertness of the thermocouple material and whether it is magnetic or not. Standard thermocouple types are listed below with the positive electrode (assuming $T_{sense} > T_{ref}$) first, followed by the negative electrode.

Nickel-alloy Thermocouples

- Type E: Type E (chromel–constantan) has a high output (68 µV/°C), which makes it well suited to cryogenic use. Additionally, it is non-magnetic. Wide range is −50 °C to +740 °C and narrow range is −110 °C to +140 °C.

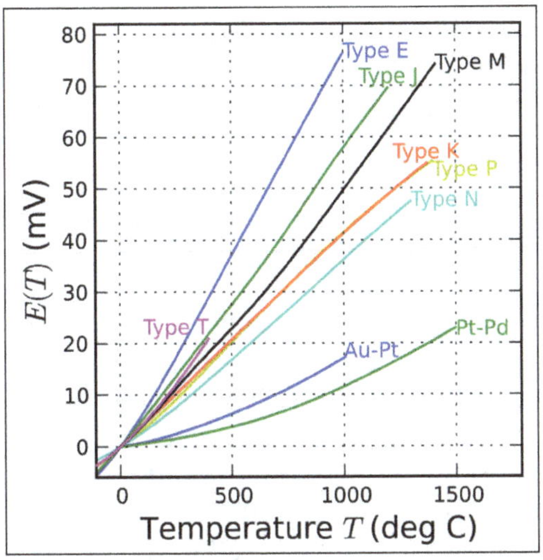

Characteristic functions for thermocouples that reach intermediate temperatures, as covered by nickel-alloy thermocouple types E, J, K, M, N, T. Also shown are the noble-metal alloy type P and the pure noble-metal combinations gold–platinum and platinum–palladium.

- Type J: Type J (iron–constantan) has a more restricted range (−40 °C to +750 °C) than type K but higher sensitivity of about 50 μV/°C. The Curie point of the iron (770 °C) causes a smooth change in the characteristic, which determines the upper temperature limit. Note, the European/German Type L is a variant of the type J, with a different specification for the EMF output.

- Type K: Type K (chromel–alumel) is the most common general-purpose thermocouple with a sensitivity of approximately 41 μV/°C. It is inexpensive, and a wide variety of probes are available in its −200 °C to +1350 °C (−330 °F to +2460 °F) range. Type K was specified at a time when metallurgy was less advanced than it is today, and consequently characteristics may vary considerably between samples. One of the constituent metals, nickel, is magnetic; a characteristic of thermocouples made with magnetic material is that they undergo a deviation in output when the material reaches its Curie point, which occurs for type K thermocouples at around 185 °C. They operate very well in oxidizing atmospheres. If, however, a mostly reducing atmosphere (such as hydrogen with a small amount of oxygen) comes into contact with the wires, the chromium in the chromel alloy oxidizes. This reduces the emf output, and the thermocouple reads low. This phenomenon is known as green rot, due to the color of the affected alloy. Although not always distinctively green, the chromel wire will develop a mottled silvery skin and become magnetic. An easy way to check for this problem is to see whether the two wires are magnetic (normally, chromel is non-magnetic).

Hydrogen in the atmosphere is the usual cause of green rot. At high temperatures, it can diffuse through solid metals or an intact metal thermowell. Even a sheath of magnesium oxide insulating the thermocouple will not keep the hydrogen out.

- Type M: Type M (82%Ni/18%Mo–99.2%Ni/0.8%Co, by weight) are used in vacuum furnaces for the same reasons as with type C. Upper temperature is limited to 1400 °C. It is less commonly used than other types.
- Type N: Type N (Nicrosil–Nisil) thermocouples are suitable for use between −270 °C and +1300 °C, owing to its stability and oxidation resistance. Sensitivity is about 39 μV/°C at 900 °C, slightly lower compared to type K.

Designed at the Defence Science and Technology Organisation (DSTO) of Australia, by Noel A. Burley, type-N thermocouples overcome the three principal characteristic types and causes of thermoelectric instability in the standard base-metal thermoelement materials:

- A gradual and generally cumulative drift in thermal EMF on long exposure at elevated temperatures. This is observed in all base-metal thermoelement materials and is mainly due to compositional changes caused by oxidation, carburization, or neutron irradiation that can produce transmutation in nuclear reactor environments. In the case of type-K thermocouples, manganese and aluminium atoms from the KN (negative) wire migrate to the KP (positive) wire, resulting in a down-scale drift due to chemical contamination. This effect is cumulative and irreversible.
- A short-term cyclic change in thermal EMF on heating in the temperature range about 250–650 °C, which occurs in thermocouples of types K, J, T, and E. This kind of EMF instability is associated with structural changes such as magnetic short-range order in the metallurgical composition.
- A time-independent perturbation in thermal EMF in specific temperature ranges. This is due to composition-dependent magnetic transformations that perturb the thermal EMFs in type-K thermocouples in the range about 25–225 °C, and in type J above 730 °C.

The Nicrosil and Nisil thermocouple alloys show greatly enhanced thermoelectric stability relative to the other standard base-metal thermocouple alloys because their compositions substantially reduce the thermoelectric instabilities. This is achieved primarily by increasing component solute concentrations (chromium and silicon) in a base of nickel above those required to cause a transition from internal to external modes of oxidation, and by selecting solutes (silicon and magnesium) that preferentially oxidize to form a diffusion-barrier, and hence oxidation-inhibiting films.

- Type T: Type T (copper–constantan) thermocouples are suited for measurements in the −200 to 350 °C range. Often used as a differential measurement, since only copper wire touches the probes. Since both conductors are non-magnetic, there is no Curie point and thus no abrupt change in characteristics. Type-T thermocouples have a sensitivity of about 43 μV/°C. Note that copper has a much higher thermal conductivity than the alloys generally used in thermocouple constructions, and so it is necessary to exercise extra care with thermally anchoring type-T thermocouples. A similar composition is found in the obsolete Type U in the German specification DIN 43712:1985-01.

Platinum/Rhodium-alloy Thermocouples

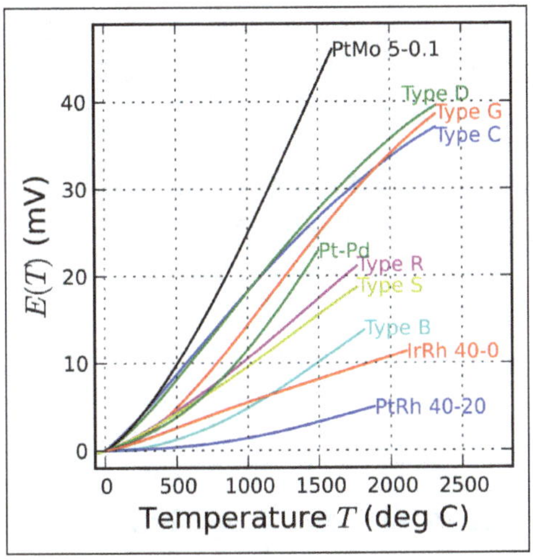

Characteristic functions for high-temperature thermocouple types, showing Pt/Rh, W/Re, Pt/Mo, and Ir/Rh-alloy thermocouples. Also shown is the Pt–Pd pure-metal thermocouple.

- Types B, R, and S thermocouples use platinum or a platinum/rhodium alloy for each conductor. These are among the most stable thermocouples, but have lower sensitivity than other types, approximately 10 μV/°C. Type B, R, and S thermocouples are usually used only for high-temperature measurements due to their high cost and low sensitivity.

- Type B: Type B (70%Pt/30%Rh–94%Pt/6%Rh, by weight) thermocouples are suited for use at up to 1800 °C. Type-B thermocouples produce the same output at 0 °C and 42 °C, limiting their use below about 50 °C. The emf function has a minimum around 21 °C, meaning that cold-junction compensation is easily performed, since the compensation voltage is essentially a constant for a reference at typical room temperatures.

- Type R: Type R (87%Pt/13%Rh–Pt, by weight) thermocouples are used 0 to 1600 °C.

- Type S: Type S (90%Pt/10%Rh–Pt, by weight) thermocouples, similar to type R, are used up to 1600 °C. Before the introduction of the International Temperature Scale of 1990 (ITS-90), precision type-S thermocouples were used as the practical standard thermometers for the range of 630 °C to 1064 °C, based on an interpolation between the freezing points of antimony, silver, and gold.

Starting with ITS-90, platinum resistance thermometers have taken over this range as standard thermometers.

Tungsten/Rhenium-alloy Thermocouples

These thermocouples are well suited for measuring extremely high temperatures. Typical uses are hydrogen and inert atmospheres, as well as vacuum furnaces. They are not used in oxidizing environments at high temperatures because of embrittlement. A typical range is 0 to 2315 °C, which can be extended to 2760 °C in inert atmosphere and to 3000 °C for brief measurements.

- Type C: (95%W/5%Re–74%W/26%Re, by weight) maximum temperature will be measured by type-c thermocouple is 2329 °C.
- Type D: (97%W/3%Re–75%W/25%Re, by weight).
- Type G: (W–74%W/26%Re, by weight).

Others Types

Chromel–gold/Iron-alloy Thermocouples

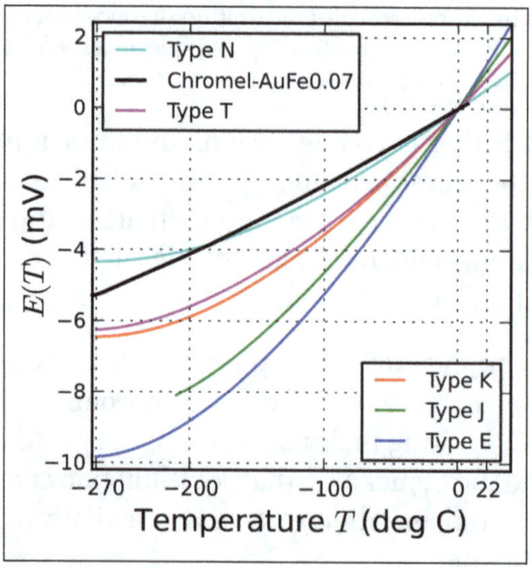

Thermocouple characteristics at low temperatures: The AuFe-based thermocouple shows a steady sensitivity down to low temperatures, whereas conventional types soon flatten out and lose sensitivity at low temperature.

In these thermocouples (chromel–gold/iron alloy), the negative wire is gold with a small fraction (0.03–0.15 atom percent) of iron. The impure gold wire gives the thermocouple a high sensitivity at low temperatures (compared to other thermocouples at that temperature), whereas the chromel wire maintains the sensitivity near room temperature. It can be used for cryogenic applications (1.2–300 K and even up to 600 K). Both the sensitivity and the temperature range depend on the iron concentration. The sensitivity is typically around 15 μV/K at low temperatures, and the lowest usable temperature varies between 1.2 and 4.2 K.

- Type P (noble-metal alloy) or "Platinel II": Type P (55%Pd/31%Pt/14% Au–65%Au/35%Pd, by weight) thermocouples give a thermoelectric voltage that mimics the type K over the range 500 °C to 1400 °C, however they are constructed purely of noble metals and so shows enhanced corrosion resistance. This combination is also known as Platinel II.

- Platinum/molybdenum-alloy thermocouples: Thermocouples of platinum/molybdenum-alloy (95%Pt/5%Mo–99.9%Pt/0.1%Mo, by weight) are sometimes used in nuclear reactors, since they show a low drift from nuclear transmutation induced by neutron irradiation, compared to the platinum/rhodium-alloy types.

- Iridium/rhodium alloy thermocouples: The use of two wires of iridium/rhodium alloys can provide a thermocouple that can be used up to about 2000 °C in inert atmospheres.

- Pure noble-metal thermocouples Au–Pt, Pt–Pd: Thermocouples made from two different, high-purity noble metals can show high accuracy even when uncalibrated, as well as low levels of drift. Two combinations in use are gold–platinum and platinum–palladium. Their main limitations are the low melting points of the metals involved (1064 °C for gold and 1555 °C for palladium). These thermocouples tend to be more accurate than type S, and due to their economy and simplicity are even regarded as competitive alternatives to the platinum resistance thermometers that are normally used as standard thermometers.

- HTIR-TC (High Temperature Irradiation Resistant) thermocouples: HTIR-TC offers a breakthrough in measuring high-temperature processes. Its characteristics are: durable and reliable at high temperatures, up to at least 1700 °C; resistant to irradiation; moderately priced; available in a variety of configurations - adaptable to each application; easily installed. Originally developed for use in nuclear test reactors, HTIR-TC may enhance the safety of operations in future reactors. This thermocouple was developed by researchers at the Idaho National Laboratory (INL).

Thermocouple Insulation

The wires that make up the thermocouple must be insulated from each other everywhere, except at the sensing junction. Any additional electrical contact between the wires, or contact of a wire to other conductive objects, can modify the voltage and give a false reading of temperature.

Typical low cost type K thermocouple (with standard type K connector). While the wires can survive and function at high temperatures, the plastic insulation will start to break down at 300 °C.

Plastics are suitable insulators for low temperatures parts of a thermocouple, whereas ceramic insulation can be used up to around 1000 °C. Other concerns (abrasion and chemical resistance) also affect the suitability of materials.

When wire insulation disintegrates, it can result in an unintended electrical contact at a different location from the desired sensing point. If such a damaged thermocouple is used in the closed loop control of a thermostat or other temperature controller, this can lead to a runaway overheating event and possibly severe damage, as the false temperature reading will typically be lower than the sensing junction temperature. Failed insulation will also typically outgas, which can lead to process contamination. For parts of thermocouples used at very high temperatures or in contamination-sensitive applications, the only suitable insulation may be vacuum or inert gas; the mechanical rigidity of the thermocouple wires is used to keep them separated.

Applications

Thermocouples are suitable for measuring over a large temperature range, from −270 up to 3000 °C (for a short time, in inert atmosphere). Applications include temperature measurement for kilns, gas turbine exhaust, diesel engines, other industrial processes and fog machines. They are less suitable for applications where smaller temperature differences need to be measured with high accuracy, for example the range 0–100 °C with 0.1 °C accuracy. For such applications thermistors, silicon bandgap temperature sensors and resistance thermometers are more suitable.

Steel Industry

Type B, S, R and K thermocouples are used extensively in the steel and iron industries to monitor temperatures and chemistry throughout the steel making process. Disposable, immersible, type S thermocouples are regularly used in the electric arc furnace process to accurately measure the temperature of steel before tapping. The cooling curve of a small steel sample can be analyzed and used to estimate the carbon content of molten steel.

Gas Appliance Safety

Many gas-fed heating appliances such as ovens and water heaters make use of a pilot flame to ignite the main gas burner when required. If the pilot flame goes out, unburned gas may be released, which is an explosion risk and a health hazard. To prevent this, some appliances use a thermocouple in a fail-safe circuit to sense when the pilot light is burning. The tip of the thermocouple is placed in the pilot flame, generating a voltage which operates the supply valve which feeds gas to the pilot. So long as the pilot flame remains lit, the thermocouple remains hot, and the pilot gas valve is held open. If the pilot light goes out, the thermocouple temperature falls, causing the voltage across the thermocouple to drop and the valve to close.

A thermocouple (the right most tube) inside the burner assembly of a water heater.

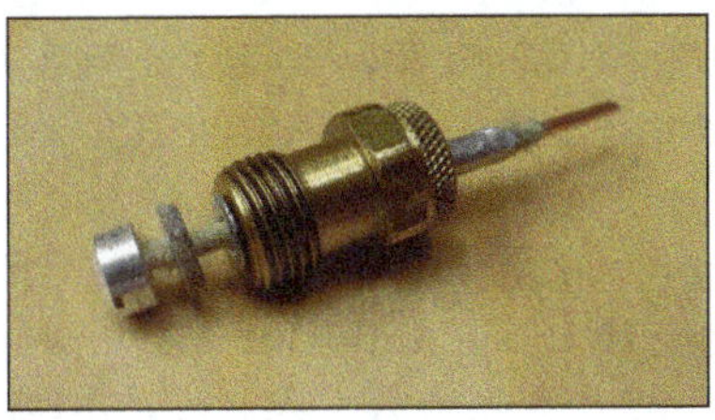

Thermocouple connection in gas appliances: The end ball (contact) on the left is insulated from the fitting by an insulating washer. The thermocouple line consists of copper wire, insulator and outer metal (usually copper) sheath which is also used as ground.

Where the probe may be easily placed above the flame, a rectifying sensor may often be used instead. With part ceramic construction, they may also be known as flame rods, flame sensors or flame detection electrodes.

Some combined main burner and pilot gas valves (mainly by Honeywell) reduce the power demand to within the range of a single universal thermocouple heated by a pilot (25 mV open circuit falling by half with the coil connected to a 10–12 mV, 0.2–0.25 A source, typically) by sizing the coil to be able to hold the valve open against a light spring, but only after the initial turning-on force is provided by the user pressing and holding a knob to compress the spring during lighting of the pilot. These systems are identifiable by the "press and hold for x minutes" in the pilot lighting instructions. (The holding current requirement of such a valve is much less than a bigger solenoid designed for pulling the valve in from a closed position would require). Special test sets are made to confirm the valve let-go and holding currents, because an ordinary milliammeter cannot be used as it introduces more resistance than the gas valve coil. Apart from testing the open circuit voltage of the thermocouple, and the near short-circuit DC continuity through the thermocouple gas valve coil, the easiest non-specialist test is substitution of a known good gas valve.

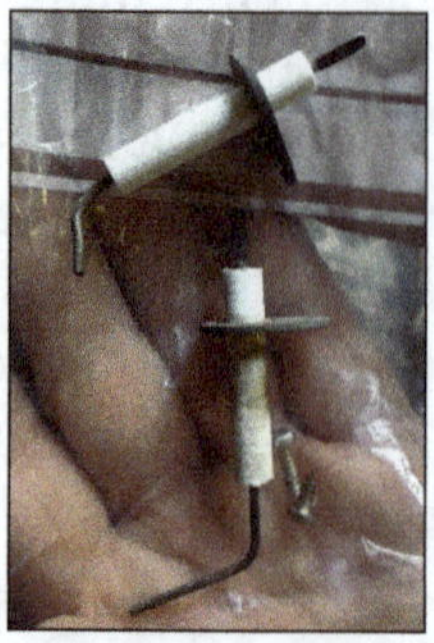

Flame-igniter(top)-and-flame-sensor.

Some systems, known as millivolt control systems, extend the thermocouple concept to both open and close the main gas valve as well. Not only does the voltage created by the pilot thermocouple activate the pilot gas valve, it is also routed through a thermostat to power the main gas valve as well. Here, a larger voltage is needed than in a pilot flame safety system, and a thermopile is used rather than a single thermocouple. Such a system requires no external source of electricity for its operation and thus can operate during a power failure, provided that all the other related system components allow for this. This excludes common forced air furnaces because external electrical power is required to operate the blower motor, but this feature is especially useful for un-powered convection heaters. A similar gas shut-off safety mechanism using a thermocouple is sometimes employed to ensure that the main burner ignites within a certain time period, shutting off the main burner gas supply valve should that not happen.

Out of concern about energy wasted by the standing pilot flame, designers of many newer appliances have switched to an electronically controlled pilot-less ignition, also called intermittent ignition. With no standing pilot flame, there is no risk of gas buildup should the flame go out, so these appliances do not need thermocouple-based pilot safety switches. As these designs lose the benefit of operation without a continuous source of electricity, standing pilots are still used in some appliances. The exception is later model instantaneous (aka “tankless”) water heaters that use the flow of water to generate the current required to ignite the gas burner; these designs also use a thermocouple as a safety cut-off device in the event the gas fails to ignite, or if the flame is extinguished.

Thermopile Radiation Sensors

Thermopiles are used for measuring the intensity of incident radiation, typically visible or infrared light, which heats the hot junctions, while the cold junctions are on a heat sink. It is possible to measure radiative intensities of only a few $\mu W/cm^2$ with commercially available thermopile sensors. For example, some laser power meters are based on such sensors; these are specifically known as thermopile laser sensor.

The principle of operation of a thermopile sensor is distinct from that of a bolometer, as the latter relies on a change in resistance.

Manufacturing

Thermocouples can generally be used in the testing of prototype electrical and mechanical apparatus. For example, switchgear under test for its current carrying capacity may have thermocouples installed and monitored during a heat run test, to confirm that the temperature rise at rated current does not exceed designed limits.

Power Production

A thermocouple can produce current to drive some processes directly, without the need for extra circuitry and power sources. For example, the power from a thermocouple can activate a valve when a temperature difference arises. The electrical energy generated by a thermocouple is converted from the heat which must be supplied to the hot side to maintain the electric potential. A continuous transfer of heat is necessary because the current flowing through the thermocouple tends to cause the hot side to cool down and the cold side to heat up (the Peltier effect).

Thermocouples can be connected in series to form a thermopile, where all the hot junctions are exposed to a higher temperature and all the cold junctions to a lower temperature. The output is the sum of the voltages across the individual junctions, giving larger voltage and power output. In a radioisotope thermoelectric generator, the radioactive decay of transuranic elements as a heat source has been used to power spacecraft on missions too far from the Sun to use solar power. Thermopiles heated by kerosene lamps were used to run batteryless radio receivers in isolated areas. There are commercially produced lanterns that use the heat from a candle to run several light-emitting diodes, and thermoelectrically-powered fans to improve air circulation and heat distribution in wood stoves.

Process Plants

Chemical production and petroleum refineries will usually employ computers for logging and for limit testing the many temperatures associated with a process, typically numbering in the hundreds. For such cases, a number of thermocouple leads will be brought to a common reference block (a large block of copper) containing the second thermocouple of each circuit. The temperature of the block is in turn measured by a thermistor. Simple computations are used to determine the temperature at each measured location.

Thermocouple as Vacuum Gauge

A thermocouple can be used as a vacuum gauge over the range of approximately 0.001 to 1 torr absolute pressure. In this pressure range, the mean free path of the gas is comparable to the dimensions of the vacuum chamber, and the flow regime is neither purely viscous nor purely molecular. In this configuration, the thermocouple junction is attached to the centre of a short heating wire, which is usually energised by a constant current of about 5 mA, and the heat is removed at a rate related to the thermal conductivity of the gas.

The temperature detected at the thermocouple junction depends on the thermal conductivity of the surrounding gas, which depends on the pressure of the gas. The potential difference measured by a thermocouple is proportional to the square of pressure over the low- to medium-vacuum range. At higher (viscous flow) and lower (molecular flow) pressures, the thermal conductivity of air or any other gas is essentially independent of pressure. The thermocouple was first used as a vacuum gauge by Voege in 1906. The mathematical model for the thermocouple as a vacuum gauge is quite complicated, as explained in detail by Van Atta, but can be simplified to:

$$P = \frac{B(V^2 - V_0^2)}{V_0^2},$$

where P is the gas pressure, B is a constant that depends on the thermocouple temperature, the gas composition and the vacuum-chamber geometry, V_0 is the thermocouple voltage at zero pressure (absolute), and V is the voltage indicated by the thermocouple.

The alternative is the Pirani gauge, which operates in a similar way, over approximately the same pressure range, but is only a 2-terminal device, sensing the change in resistance with temperature of a thin electrically heated wire, rather than using a thermocouple.

Piezoelectric Transducer (Sensor)

A piezoelectric sensor is a device that uses the piezoelectric effect to measure changes in pressure, acceleration, temperature, strain, or force by converting them to an electrical charge.

Applications

Piezoelectric sensors are versatile tools for the measurement of various processes. They are used for quality assurance, process control, and for research and development in many industries. Pierre Curie discovered the piezoelectric effect in 1880, but only in the 1950s did manufacturers begin to use the piezoelectric effect in industrial sensing applications. Since then, this measuring principle has been increasingly used, and has become a mature technology with excellent inherent reliability.

They have been successfully used in various applications, such as in medical, aerospace, nuclear instrumentation, and as a tilt sensor in consumer electronics or a pressure sensor in the touch pads of mobile phones. In the automotive industry, piezoelectric elements are used to monitor combustion when developing internal combustion engines.

The sensors are either directly mounted into additional holes into the cylinder head or the spark/glow plug is equipped with a built-in miniature piezoelectric sensor.

The rise of piezoelectric technology is directly related to a set of inherent advantages. The high modulus of elasticity of many piezoelectric materials is comparable to that of many metals and goes up to 10^6 N/m². Even though piezoelectric sensors are electromechanical systems that react to compression, the sensing elements show almost zero deflection. This gives piezoelectric sensors ruggedness, an extremely high natural frequency and an excellent linearity over a wide amplitude range. Additionally, piezoelectric technology is insensitive to electromagnetic fields and radiation, enabling measurements under harsh conditions. Some materials used (especially gallium phosphate or tourmaline) are extremely stable at high temperatures, enabling sensors to have a working range of up to 1000 °C. Tourmaline shows pyroelectricity in addition to the piezoelectric effect; this is the ability to generate an electrical signal when the temperature of the crystal changes. This effect is also common to piezoceramic materials. Gautschi in Piezoelectric Sensorics offers this comparison table of characteristics of piezo sensor materials vs other types:

Principle	Strain Sensitivity [V/με]	Threshold [με]	Span to threshold ratio
Piezoelectric	5.0	0.00001	100,000,000
Piezoresistive	0.0001	0.0001	2,500,000
Inductive	0.001	0.0005	2,000,000
Capacitive	0.005	0.0001	750,000
Resistive	0.000005	0.01	50,000

One disadvantage of piezoelectric sensors is that they cannot be used for truly static measurements. A static force results in a fixed amount of charge on the piezoelectric material. In conventional readout electronics, imperfect insulating materials and reduction in internal sensor resistance causes a constant loss of electrons and yields a decreasing signal. Elevated temperatures cause an additional drop in internal resistance and sensitivity. The main effect on the piezoelectric effect is that with increasing pressure loads and temperature, the sensitivity reduces due to twin formation. While quartz sensors must be cooled during measurements at temperatures above 300 °C, special types of crystals like GaPO4 gallium phosphate show no twin formation up to the melting point of the material itself.

However, it is not true that piezoelectric sensors can only be used for very fast processes or at ambient conditions. In fact, numerous piezoelectric applications produce quasi-static measurements, and other applications work in temperatures higher than 500 °C.

Piezoelectric sensors can also be used to determine aromas in the air by simultaneously measuring resonance and capacitance. Computer controlled electronics vastly increase the range of potential applications for piezoelectric sensors.

Piezoelectric sensors are also seen in nature. The collagen in bone is piezoelectric, and is thought by some to act as a biological force sensor.

Principle of Operation

The way a piezoelectric material is cut defines one of its three main operational modes:

- Transverse.
- Longitudinal.
- Shear.

Transverse Effect

A force applied along a neutral axis (y) displaces charges along the (x) direction, perpendicular to the line of force. The amount of charge (Q_x) depends on the geometrical dimensions of the respective piezoelectric element. When dimensions a, b, d apply,

$$Q_x = d_{xy} F_y b / a,$$

where a is the dimension in line with the neutral axis, b is in line with the charge generating axis and d is the corresponding piezoelectric coefficient.

Longitudinal Effect

The amount of charge displaced is strictly proportional to the applied force and independent of the piezoelectric element size and shape. Putting several elements mechanically in series and electrically in parallel is the only way to increase the charge output. The resulting charge is,

$$Q_x = d_{xx} F_x n,$$

where d_{xx} is the piezoelectric coefficient for a charge in x-direction released by forces applied along x-direction (in pC/N). F_x is the applied Force in x-direction [N] and n corresponds to the number of stacked elements.

Shear Effect

The charge produced is exactly proportional to the applied force and is generated at a right angle to the force. The charge is independent of the element size and shape. For n elements mechanically in series and electrically in parallel the charge is,

$$Q_x = 2 d_{xx} F_x n.$$

In contrast to the longitudinal and shear effects, the transverse effect make it possible to fine-tune sensitivity on the applied force and element dimension.

Electrical Properties

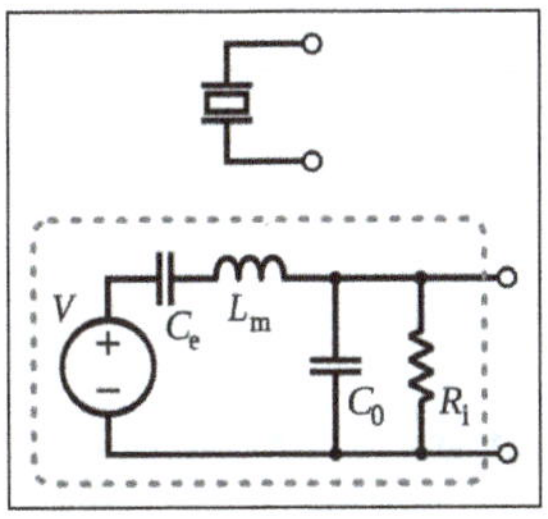

Schematic symbol and electronic model of a piezoelectric sensor.

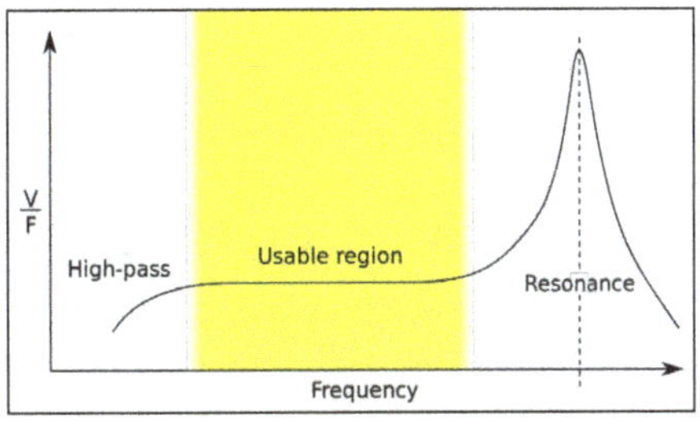

Frequency response of a piezoelectric sensor; output voltage vs applied force.

A piezoelectric transducer has very high DC output impedance and can be modeled as a proportional voltage source and filter network. The voltage *V* at the source is directly proportional to the applied force, pressure, or strain. The output signal is then related to this mechanical force as if it had passed through the equivalent circuit.

A detailed model includes the effects of the sensor's mechanical construction and other non-idealities. The inductance L_m is due to the seismic mass and inertia of the sensor itself. C_e is inversely proportional to the mechanical elasticity of the sensor. C_o represents the static capacitance of the transducer, resulting from an inertial mass of infinite size. R_i is the insulation leakage resistance of the transducer element. If the sensor is connected to a load resistance, this also acts in parallel with the insulation resistance, both increasing the high-pass cutoff frequency.

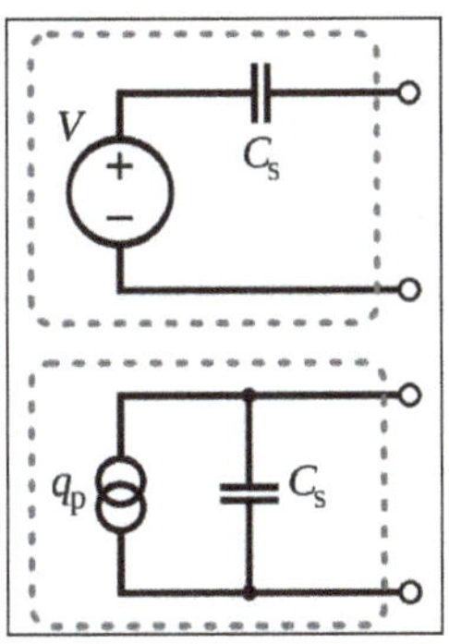

In the flat region, the sensor can be modeled as a voltage source in series with the sensor's capacitance or a charge source in parallel with the capacitance.

For use as a sensor, the flat region of the frequency response plot is typically used, between the high-pass cutoff and the resonant peak. The load and leakage resistance must be large enough that low frequencies of interest are not lost. A simplified equivalent circuit model can be used in this region, in which C_s represents the capacitance of the sensor surface itself, determined by the standard formula for capacitance of parallel plates. It can also be modeled as a charge source in parallel with the source capacitance, with the charge directly proportional to the applied force.

Sensor Design

Metal disks with piezo material, used in buzzers or as contact microphones.

Based on piezoelectric technology various physical quantities can be measured the most common are pressure and acceleration. For pressure sensors, a thin membrane and a massive base is used, ensuring that an applied pressure specifically loads the elements in one direction. For accelerometers, a seismic mass is attached to the crystal elements. When the accelerometer experiences a motion, the invariant seismic mass loads the elements according to Newton's second law of motion $F = ma$.

The main difference in working principle between these two cases is the way they apply forces to the sensing elements. In a pressure sensor, a thin membrane transfers the force to the elements, while in accelerometers an attached seismic mass applies the forces. Sensors often tend to be sensitive to more than one physical quantity. Pressure sensors show false signal when they are exposed to vibrations. Sophisticated pressure sensors therefore use acceleration compensation elements in addition to the pressure sensing elements. By carefully matching those elements, the acceleration signal (released from the compensation element) is subtracted from the combined signal of pressure and acceleration to derive the true pressure information.

Vibration sensors can also harvest otherwise wasted energy from mechanical vibrations. This is accomplished by using piezoelectric materials to convert mechanical strain into usable electrical energy.

Sensing Materials

Two main groups of materials are used for piezoelectric sensors: piezoelectric ceramics and single crystal materials. The ceramic materials (such as PZT ceramic) have a piezoelectric constant/sensitivity that is roughly two orders of magnitude higher than those of the natural single crystal materials and can be produced by inexpensive sintering processes. The piezoeffect in piezoceramics is "trained", so their high sensitivity degrades over time. This degradation is highly correlated with increased temperature.

The less-sensitive, natural, single-crystal materials (gallium phosphate, quartz, tourmaline) have a higher – when carefully handled, almost unlimited – long term stability.

There are also new single-crystal materials commercially available such as Lead Magnesium Niobate-Lead Titanate (PMN-PT). These materials offer improved sensitivity over PZT but have a lower maximum operating temperature and are currently more expensive to manufacture.

Passive Transducer

Resistive Transducer

The transducer whose resistance varies because of the environmental effects such type of transducer is known as the resistive transducer. The change in resistance is measured by the ac or dc measuring devices. The resistive transducer is used for measuring the physical quantities like temperature, displacement, vibration etc.

The measurement of the physical quantity is quite difficult. The resistive transducer converts the physical quantities into variable resistance which is easily measured by the meters. The process of variation in resistance is widely used in the industrial applications.

The resistive transducer can work both as the primary as well as the secondary transducer. The primary transducer changes the physical quantities into a mechanical signal, and secondary transducer directly transforms it into an electrical signal.

Example: The circuit of the sliding resistive transducer is shown in the figure below. The sliding contacts are placed on the resistive element. The slider moves horizontally. The movement of the slider changes the value of the resistive element of the transducer which is measured by the voltage source E.

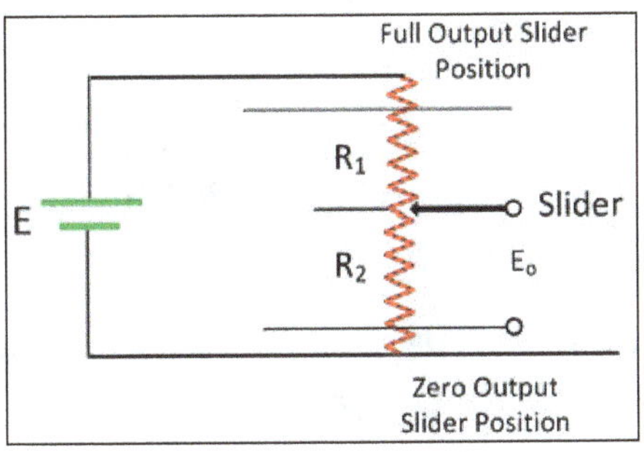

Sliding Resistive Transducer.

The displacement of the slider is converted into an electrical signal.

Advantages of Resistive Transducer

The following are the advantages of the resistive transducer:

- Both the AC and DC, current or voltage is appropriate for the measurement of variable resistance.

- The resistive transducer gives the fast response.
- It is available in various sizes and having a high range of resistance.

Working Principle of Resistive Transducer

The resistive transducer element works on the principle that the resistance of the element is directly proportional to the length of the conductor and inversely proportional to the area of the conductor.

$$R = \rho L / A$$

where, R = Resistance in ohms.

A = Cross-section area of the conductor in meter square.

L = Length of the conductor in meter square.

ρ = The resistivity of the conductor in materials in ohm meter.

The resistive transducer is designed by considering the variation of the length, area and resistivity of the metal.

Applications of Resistive Transducer

The following are the applications of the resistive transducer:

- Potentiometer: The translation and rotatory potentiometer are the examples of the resistive transducers. The resistance of their conductor varies with the variation in their lengths which is used for the measurement of displacement.
- Strain gauges: The resistance of their semiconductor material changes when the strain occurs on it. This property of metals is used for the measurement of the pressure, force-displacement etc.
- Resistance Thermometer: The resistance of the metals changes because of changes in temperature. This property of conductor is used for measuring the temperature.
- Thermistor: It works on the principle that the temperature coefficient of the thermistor material varies with the temperature. The thermistor has the negative temperature coefficient. The Negative temperature coefficient means the temperature is inversely proportional to resistance.

There are a number of ways because of which the resistance of the metal changes with the changed in the physical phenomenon. And this property of conductors is used for measuring the physical quantities of material.

Inductive Transducers

The inductive transducers work on the principle of the magnetic induction of magnetic material. Just as the resistance of the electric conductor depends on number of factors, the induction of the magnetic material depends on a number of variables like the number of turns of the coil on the material, the size of the magnetic material, and the permeability of the flux path. In the inductive transducers the magnetic materials are used in the flux path and there are one or more air gaps. The change in the air gap also results in change in the inductance of the circuit and in most of the inductive transducers it is used for the working of the instrument.

There are two common type inductive transducers: simple inductance type and two-coil mutual inductance type. Both have been described below along with their circuits.

Simple Inductance Type Inductive Transducers

In the simple inductance type of the inductive transducers simple single coil is used as the transducer. When the mechanical element whose displacement is to be measured is moved, it changes the permeance of the flux path generated by the circuit, which changes the inductance of the circuit and the corresponding output. The output from the circuit is calibrated directly against the value of the input, thus it directly gives the valve of the parameter to be measured.

The figure below shows the single coil inductive circuit. Here the magnetic material is connected to the electric circuit and it is excited by the alternating current. At the bottom there is another magnetic material that acts as the armature. As the armature is moved, the air gap between the two magnetic material changes and the permeance of the flux generated by the circuit changes that changes the inductance of the circuit and its output. The output meter directly gives the valve of the input mechanical quantity.

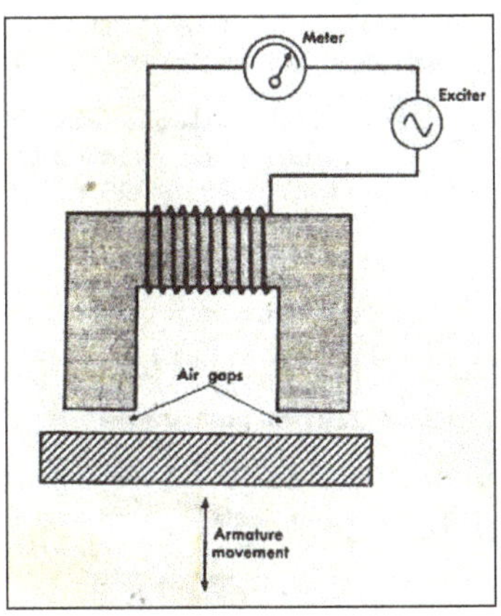

A simple self inductance arrangement where in a change in the air gap changes the pickup output.

In the figure, coil is wound around the round hollow magnetic material and there is magnetic core that moves inside hollow magnetic material. In the above circuits the change in the air gap or the change in the amount of the magnetic material in the circuit can be used to produce the output proportional to the input.

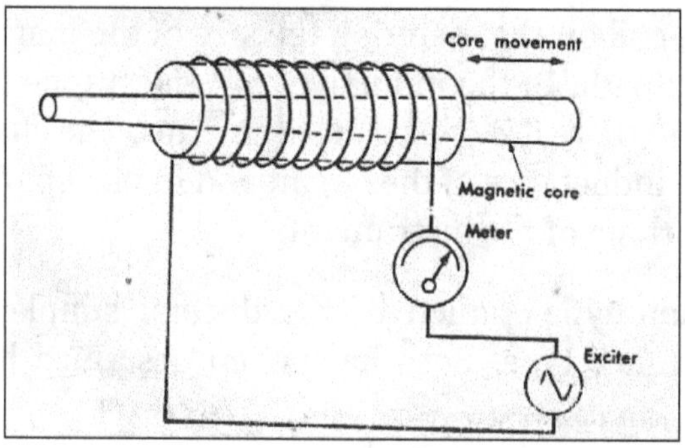

Another arrangement of the coils is shown in figure below, where two coils are used. In this circuit the movement of the core changes the relative inductance of the two coils and over all inductance of the circuit. This system is used in the devices along with the inductive bridge circuit. In this circuit the change in the induction ratio of the two coils provides the output proportional to the mechanical input.

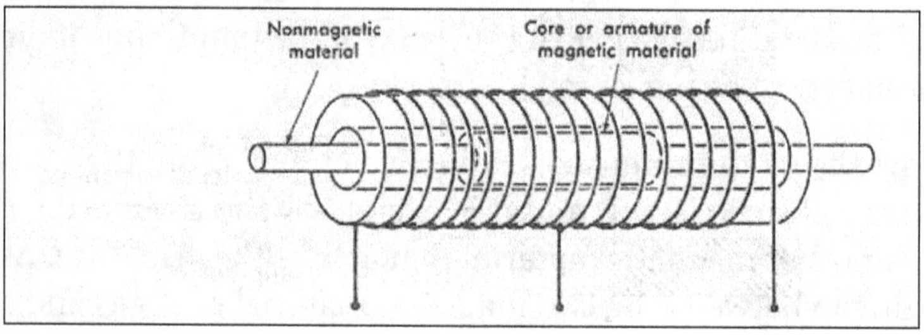

Capacitive Transducer

The capacitive transducer is used for measuring the displacement, pressure and other physical quantities. It is a passive transducer that means it requires external power for operation. The capacitive transducer works on the principle of variable capacitances. The capacitance of the capacitive transducer changes because of many reasons like overlapping of plates, change in distance between the plates and dielectric constant.

The capacitive transducer contains two parallel metal plates. These plates are separated by the dielectric medium which is either air, material, gas or liquid. In the normal capacitor the distance between the plates are fixed, but in capacitive transducer the distance between them are varied.

The capacitive transducer uses the electrical quantity of capacitance for converting the mechanical movement into an electrical signal. The input quantity causes the change of the capacitance which is directly measured by the capacitive transducer. The capacitors measure both the static and dynamic changes. The displacement is also measured directly by connecting the measurable devices to the movable plate of the capacitor. It works on with both the contacting and non-contacting modes.

Principle of Operation

The equations below express the capacitance between the plates of a capacitor:

$$C = \varepsilon A / d$$
$$C = \varepsilon_r \varepsilon_0 A / d$$

where, A = Overlapping area of plates in m^2.

d = The distance between two plates in meter.

ε = Permittivity of the medium in F/m.

ε_r = Relative permittivity.

ε_o = The permittivity of free space.

The schematic diagram of a parallel plate capacitive transducer is shown in the figure below:

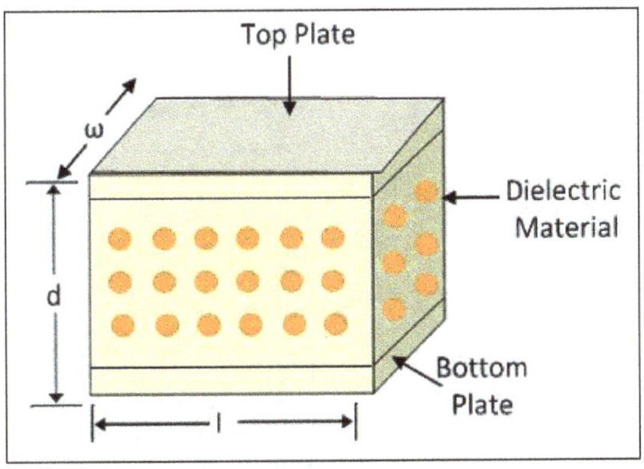

Parallel Plate Capacitive Transducer.

The change in capacitance occurs because of the physicals variables like displacement, force, pressure, etc. The capacitance of the transducer also changes by the variation in their dielectric constant which is usually because of the measurement of liquid or gas level.

The capacitance of the transducer is measured with the bridge circuit. The output impedance of transducer is given as,

$$X_c = 1 / 2\pi f c$$

Where, C = Capacitance.

f = Frequency of excitation in Hz.

The capacitive transducer is mainly used for measurement of linear displacement. The capacitive transducer uses the following three effects:

- Variation in capacitance of transducer is because of the overlapping of capacitor plates.
- The change in capacitance is because of the change in distances between the plates.
- The capacitance changes because of dielectric constant.

The following methods are used for the measuring displacement:

A Transducer using the Change in the Area of Plates

The equation below shows that the capacitance is directly proportional to the area of the plates. The capacitance changes correspondingly with the change in the position of the plates.

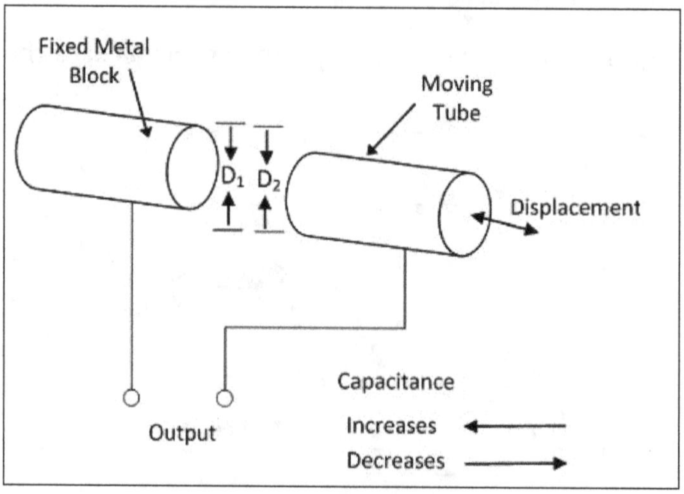

Capacitive Transducer.

The capacitive transducers are used for measuring the large displacement approximately from 1mm to several cms. The area of the capacitive transducer changes linearly with the capacitance and the displacement. Initially, the nonlinearity occurs in the system because of the edges. Otherwise, it gives the linear response.

The capacitance of the parallel plates is given as,

$$C = \frac{\varepsilon A}{d} = \frac{\varepsilon x \omega}{d} F$$

where x – the length of overlapping part of plates,

ω – the width of overlapping part of plates.

The sensitivity of the displacement is constant, and therefore it gives the linear relation between the capacitance and displacement.

$$S = \frac{\partial C}{\partial x} = \varepsilon \frac{\omega}{d} F/m$$

The capacitive transducer is used for measuring the angular displacement. It is measured by the movable plates shown below. One of the plates of the transducer is fixed, and the other is movable.

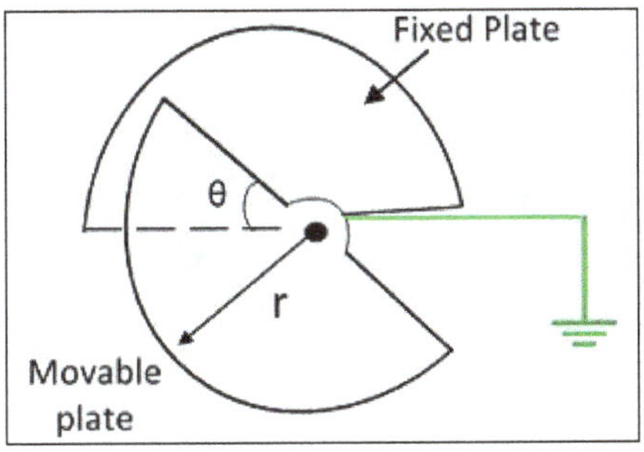

The phasor diagram of the transducer is shown in the figure below:

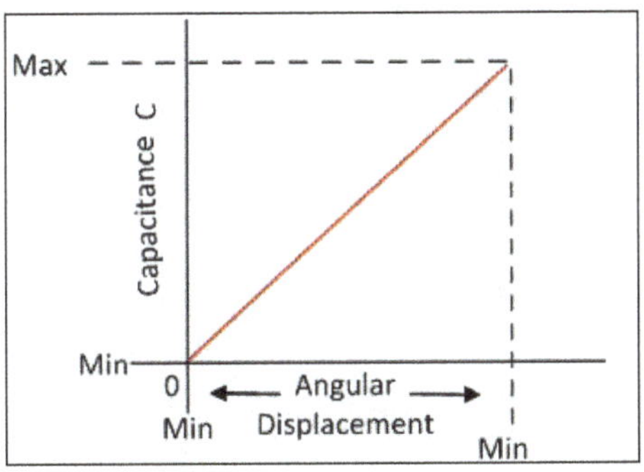

The angular movement changes the capacitance of the transducers. The capacitance between them is maximum when these plates overlap each other. The maximum value of capacitance is expressed as,

$$C_{max} = \frac{\varepsilon A}{d} = \frac{\pi \varepsilon r^2}{2d}$$

The capacitance at angle θ is given expressed as,

$$C = \frac{\varepsilon \theta r^2}{2d}$$

θ – angular displacement in radian. The sensitivity for the change in capacitance is given as,

$$S = \frac{\partial C}{\partial \theta} = \frac{\varepsilon r^2}{2d}$$

The 180° is the maximum value of the angular displacement of the capacitor.

The Transducer using the Change in Distance between the Plates

The capacitance of the transducer is inversely proportional to the distance between the plates. The one plate of the transducer is fixed, and the other is movable. The displacement which is to be measured links to the movable plates.

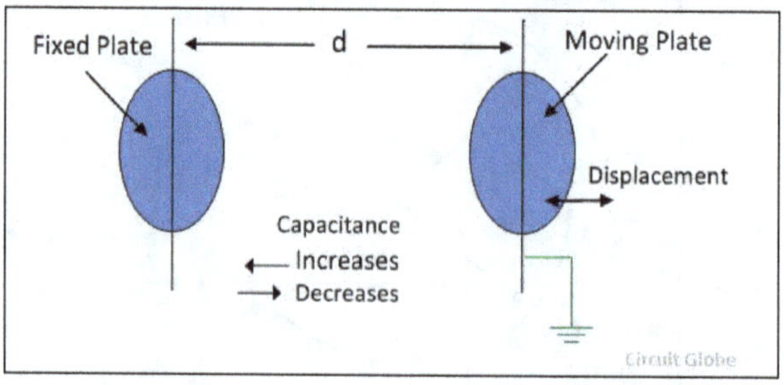

The capacitance is inversely proportional to the distance because of which the capacitor shows the nonlinear response. Such type of transducer is used for measuring the small displacement. The phasor diagram of the capacitor is shown in the figure below:

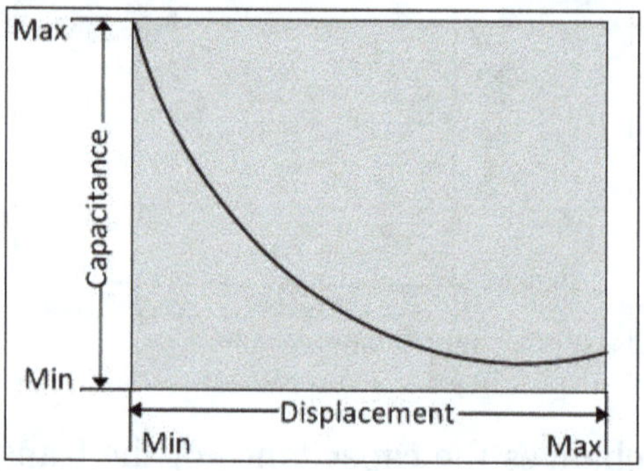

The sensitivity of the transducer is not constant and vary from places to places.

Advantage of Capacitive Transducer

The following are the major advantages of capacitive transducers:

- It requires an external force for operation and hence very useful for small systems.
- The capacitive transducer is very sensitive.
- It gives good frequency response because of which it is used for the dynamic study.

- The transducer has high input impedance hence they have a small loading effect.
- It requires small output power for operation.

Disadvantages of Capacitive Transducer

The main disadvantages of the transducer are as follows:

- The metallic parts of the transducers require insulation.
- The frame of the capacitor requires earthing for reducing the effect of the stray magnetic field.
- Sometimes the transducer shows the nonlinear behaviours because of the edge effect which is controlled by using the guard ring.
- The cable connecting across the transducer causes an error.

Uses of Capacitive Transducer

The following are the uses of capacitive transducer:

- The capacitive transducer uses for measurement of both the linear and angular displacement. It is extremely sensitive and used for the measurement of very small distance.
- It is used for the measurement of the force and pressures. The force or pressure, which is to be measured, is first converted into a displacement, and then the displacement changes the capacitances of the transducer.
- It is used as a pressure transducer in some cases, where the dielectric constant of the transducer changes with the pressure.
- The humidity in gases is measured through the capacitive transducer.
- The transducer uses the mechanical modifier for measuring the volume, density, weight etc.

The accuracy of the transducer depends on the variation of temperature to the high level.

Displacement Transducers

Linear Variable Differential Transformer

The linear variable differential transformer (LVDT) (also called linear variable displacement transformer, linear variable displacement transducer, or simply differential transformer) is a type of electrical transformer used for measuring linear displacement (position). A counterpart to this device that is used for measuring rotary displacement is called a rotary variable differential transformer (RVDT).

LVDTs are robust, absolute linear position/displacement transducers; inherently frictionless, they have a virtually infinite cycle life when properly used. As AC operated LVDTs do not contain any electronics, they can be designed to operate at cryogenic temperatures or up to 1200 °F (650 °C), in harsh environments, under high vibration and shock levels. LVDTs have been widely used in applications such as power turbines, hydraulics, automation, aircraft, satellites, nuclear reactors, and many others. These transducers have low hysteresis and excellent repeatability.

The LVDT converts a position or linear displacement from a mechanical reference (zero or null position) into a proportional electrical signal containing phase (for direction) and amplitude (for distance) information. The LVDT operation does not require an electrical contact between the moving part (probe or core assembly) and the coil assembly, but instead relies on electromagnetic coupling.

LVDT: Construction and Working Principle

Construction

LVDT is a transformer consisting of one primary winding P and two secondary winding S_1 & S_2 mounted on a cylindrical former. The two secondary winding have equal number of turns and placed identically on either side of the primary winding as shown in figure below:

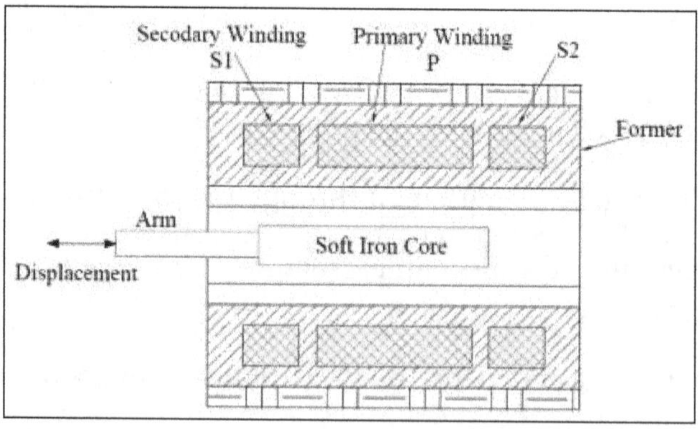

A movable soft iron core is placed inside the former. Actually the movable core is made of nickel iron with hydrogen annealed. Hydrogen annealing is done to eliminate harmonics, residual voltage of core and thus provides high sensitivity. The movable core also is laminated in order to reduce eddy current loss. The assembly of laminated core is placed in a cylindrical steel housing and end lids are provided for electromagnetic and electrostatic shielding. The displacement to be measured is attached to this movable soft iron core.

LVDT: Working Principle

Since the primary winding of Linear Variable Differential Transformer (LVDT) is supplied with AC supply, it produces an alternating magnetic flux in the core which in turn link with the secondary winding S_1 and S_2 to produce emf due to transformer action.

The electrical equivalent circuit of LVDT is shown below:

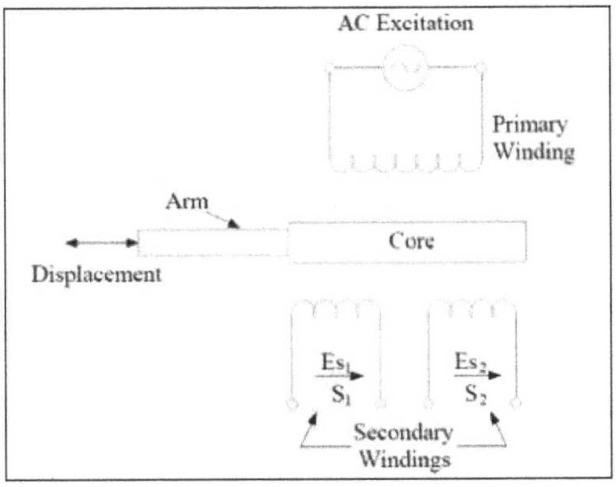

Let us assume that the emf produced in secondary winding S_1 is E_{s1} and that in S_2 is E_{s2}. The magnitude of E_{s1} and E_{s2} will depend upon the magnitude of rate of change of flux (dØ / dt) as per the Faraday's Law. The lower the value of 'dt', the more will be the emf induced. But lower value of 'dt' means that core is moving faster. Thus we can say that the faster the movement of core, the greater will be the magnitude of emf induced in secondary windings.

To get a single output voltage from the Linear Variable Differential Transformer (LVDT), both the secondary winding are connected in series but in phase opposition as shown in figure below:

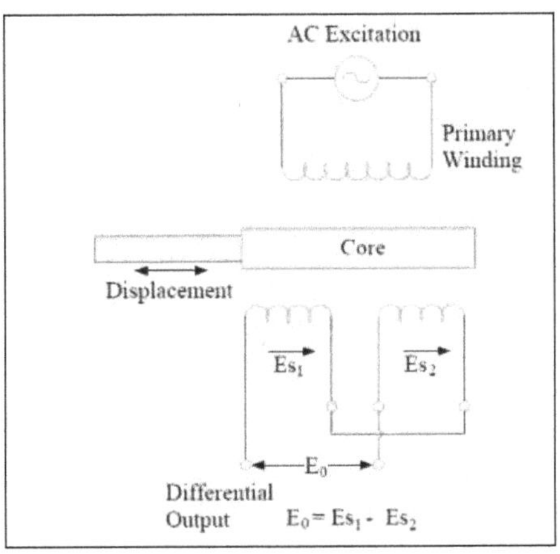

Due to this connection, the net output voltage E_o of the LVDT is given as below:

$$E_o = E_{s1} - E_{s2}$$

Since the secondary windings of LVDT are identical and placed symmetrically on either side of core, therefore under normal position the flux linkage of both the secondary winding S_1 & S_2 will be same. This means $E_{s1} = E_{s2}$ and hence net output voltage E_o of LVDT = 0. This position of soft iron core is called NULL position. Thus NULL position of Linear Variable Differential Transformer is the normal position of movable core where the net output voltage is zero.

Now, as the core can either be moved toward right or left to the null position. Let us now consider such movement of core under two cases.

Core is moved Left to the NULL Position

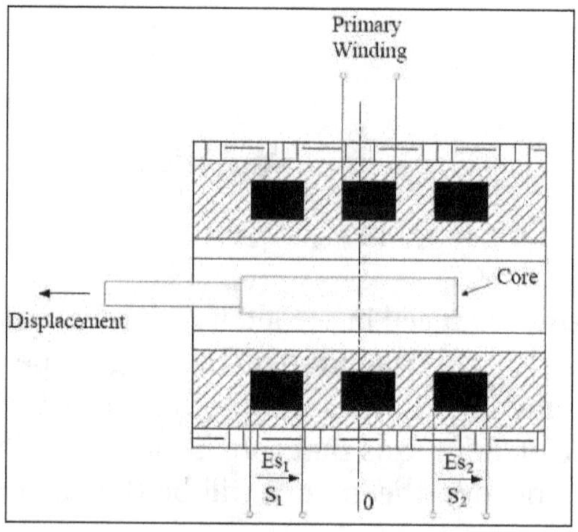

When core of LVDT is moved to the left of the NULL position 'O' as shown in figure above, the flux linkage of secondary winding S_1 will become more than that of winding S_2. This means the emf induced in winding S_1 will be more than S_2. Hence $E_{s1} > E_{s2}$ and net output voltage $E_o = (E_{s1} - E_{s2})$ = Positive. This means that the output voltage E_o will be in phase with the primary voltage.

Core is Moved Right to the NULL Position

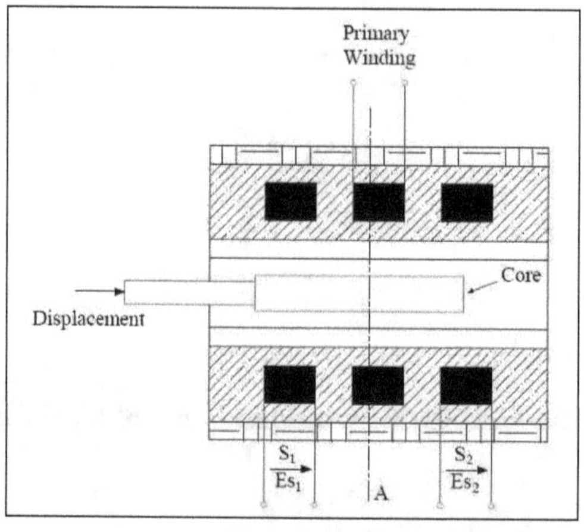

When the core of LVDT is moved toward right of NULL position 'A', you can guess what will happen? Obviously the emf induced in secondary winding S_2 will be more than that of S_1. This means $E_{s2} > E_{s1}$ and hence net output voltage $E_o = (E_{s1} - E_{s2})$ = negative. This means that the output voltage of LVDT will be in phase opposition (180 degree out of phase) with the primary voltage.

From the above two cases, we can have the following conclusions:

- The direction of movement of a physical quantity can be identified by the output voltage of LVDT. If the output voltage E_o is positive, this means the physical quantity is moving toward left.
- If the output voltage E_o is negative, this will mean that the physical quantity is moving in the right direction from the NULL position.
- The amount/magnitude of displacement is proportional to the magnitude of output voltage. The more the output voltage, the more will be displacement. But here is a clue. You can't take core out of the former; otherwise the output voltage will become zero.
- In fact corresponding to both the cases i.e. whether core is moving left or right to the NULL position, the output voltage will increase lineally up to a displacement of around 5 mm from the NULL position. After 5mm, output voltage E_o becomes non-linear. The graph of variation of E_o with displacement is shown below:

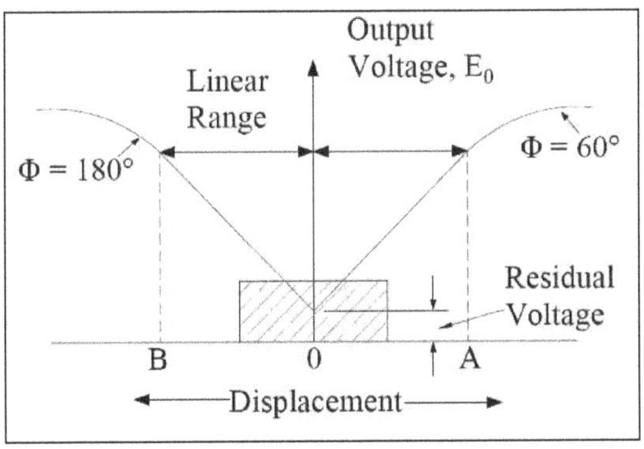

Carefully observe the above graph. It may be noted from the graph that even at NULL position (i.e. when there is no displacement) there is some output voltage of LVDT. This small output is due to the residual magnetism in the iron core.

Application

LVDT is used in those applications where displacement ranging from fraction of a mm to few cm. As a primary transducer, it converts the mechanical displacement into electrical signal. Acting as a secondary transducer, it is sued for measurement of force, pressure, weight etc.

Rotary Variable Differential Transformer

The RVDT stands for Rotary Variable Differential Transformer. It is one kind of electromechanical transducer used to give the linear o/p which is proportional to the i/p angular displacement. The main function of RVDT is to detect the angular displacement and converts it into an electrical signal. The both the RVDT and LVDT workings are similar, but LVDT employs the flexible iron core for displacement measurement whereas in RVDT employs a cam type core. This core will turn among the two windings of the transformer using the shaft.

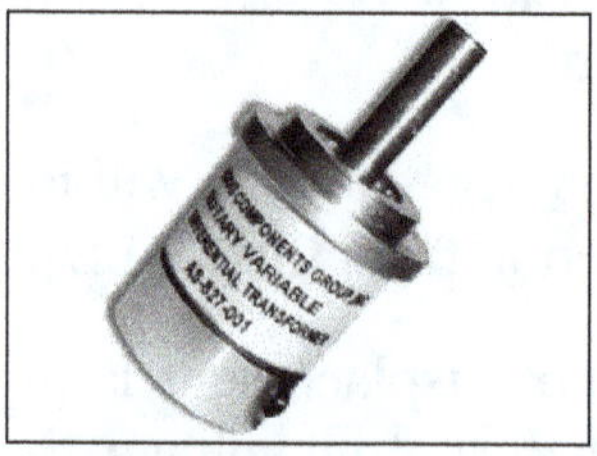

Rotary Variable Differential Transformer.

RVDT Construction and its Working

RVDT transducer has two windings similar to a normal transformer such as primary winding and two secondary windings shown in the following RVDT diagram. The two windings of the transformer wounded, where the two secondary windings have an equivalent number of windings. These are located on both sides of the primary winding of the transformer. A cam formed a magnetic core which is made with a soft iron is coupled to a shaft. Thus, this core can be twisted among the windings. The construction of both the RVDT and LVDT are similar but the main difference is the shape of the core in transformer windings. This core will turn between the two windings of the transformer due to the shaft.

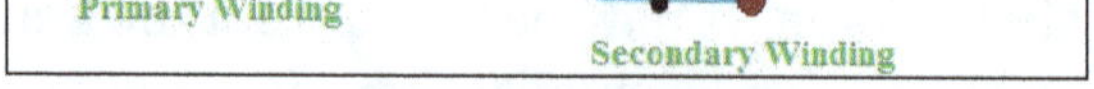

RVDT Construction.

The typical RVDTs are linear over a +40 or -40 degrees, Sensitivity is about 2mV to 3mV per degree of rotation and the input voltage range is 3V RMS at frequency ranges from 400Hz to 20kHz. Based on the movement of the shaft in the transformer, the three conditions will be produced such as:

- When the Core is at Null Position: In the first condition, when the shaft is placed at the null position then the induced e.m.f in the secondary windings are similar although reverse in phase. Thus, the differential o/p potential will be zero, and the condition will be E1 = E2, where Eo = E1-E2 =0.

- When the Core Rotates in Clockwise Direction: In the second condition, when the shaft rotates in the direction of clockwise; more section of the core will enter across the primary winding. Therefore, the induced e.m.f across the primary winding is higher than secondary winding. Hence, the differential o/p potential is positive, and the condition will be E1 > E2, where Eo = E1-E2 = positive.
- When the Core Rotates in Anticlockwise Direction: In the third condition, when the shaft rotates in the direction of anticlockwise, more section of the core will be entered across the secondary winding. Thus, the induced e.m.f across the secondary coil is higher than the primary coil. Hence, the differential o/p potential is negative that means 1800 phase shift, and the condition will be E1 < E2, where Eo = E1-E2 = negative.

RVDT Advantages and Common Applications

One variation of the linear variable differential transformer is a rotary variable differential transformer. Instead of sensing linear motion, a rotary variable differential transformer senses angular movement. It is also considered a passive transducer.

For an RVDT to produce the necessary output, the sensor produces an output voltage of variable alternating current which is proportional to the angular displacement of the input shaft. The construction of an RVDT is similar to an LVDT except for the core. Instead of three windings, an RVDT only has two which are the primary and secondary winding mechanisms.

Advantages of using an RVDT

There are several advantages to using an RVDT over other types of position sensors. Although the accuracy of an RVDT may be limited, there are notable benefits such as:

- Durability: RVDTs are durable and robust sensing mechanisms. Applications which require a sensor that can withstand extreme conditions such as vibration and shock will benefit from using an RVDT.
- Inexpensive: Cost-effectiveness is an important characteristic especially when the application has a limited budget. Although RVDTs are generally cheap to use, they are still quite robust and reliable.
- Easy to manage electronic parts: The electrical components of an RVDT are easy to operate and simple to maintain. This sensor does not contain complicated circuitry or require specialty auxiliary parts to function.
- Compact size: Another benefit of an RVDT is its relatively small size and compact design. It fits many applications where there is limited space to install a sensor.

Like an LVDT, RVDTs are some of the most commonly used sensors today. Due to its contactless construction, it does not suffer from any functional integrity issues which result from friction and wear and tear.

RVDTs have a well-established reputation for reliability even in the harshest working conditions. As such, it is a preferred sensor in manufacturing heavy duty equipment within the aerospace and oil and gas industries. In addition to these, other typical applications of RVDTs include:

- Hydraulic pumps,
- Rotary actuators,
- Valve positioning,
- Throttle lever positioning.

Most manufacturers of position sensors such as RVDTs use contactless design to maintain the sturdiness of its mechanism.

Pressure Transducer

Strain Gauge

A strain gauge is a device used to measure strain on an object. Invented by Edward E. Simmons and Arthur C. Ruge in 1938, the most common type of strain gauge consists of an insulating flexible backing which supports a metallic foil pattern. The gauge is attached to the object by a suitable adhesive, such as cyanoacrylate. As the object is deformed, the foil is deformed, causing its electrical resistance to change. This resistance change, usually measured using a Wheatstone bridge, is related to the strain by the quantity known as the gauge factor.

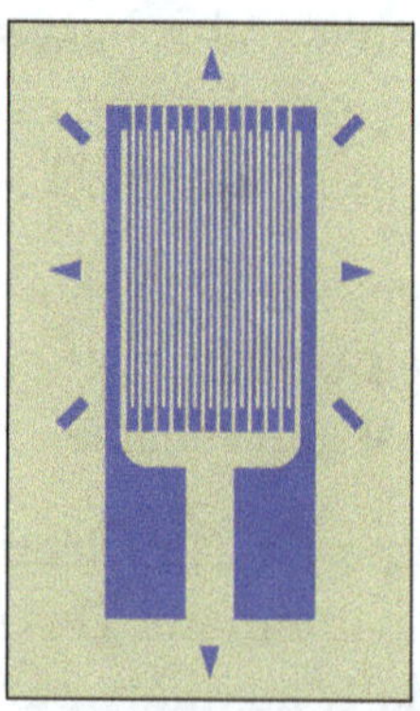

Typical foil strain gauge; the blue region is conductive and resistance is measured from one large blue pad to the other. The gauge is far more sensitive to strain in the vertical direction than in the horizontal direction. The markings outside the active area help to align the gauge during installation.

Physical Operation

A strain gauge takes advantage of the physical property of electrical conductance and its dependence on the conductor's geometry. When an electrical conductor is stretched within the limits of its elasticity such that it does not break or permanently deform, it will become narrower and longer, which increases its electrical resistance end-to-end. Conversely, when a conductor is compressed such that it does not buckle, it will broaden and shorten, which decreases its electrical resistance end-to-end. From the measured electrical resistance of the strain gauge, the amount of induced stress may be inferred.

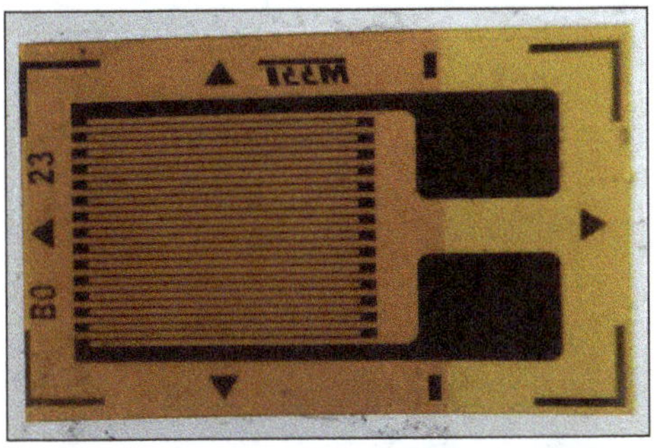

An unmounted resistive foil strain gauge.

A typical strain gauge arranges a long, thin conductive strip in a zig-zag pattern of parallel lines. This does not increase the sensitivity, since the percentage change in resistance for a given strain for the entire zig-zag is the same as for any single trace. A single linear trace would have to be extremely thin, hence liable to overheating (which would change its resistance and cause it to expand), or would need to be operated at a much lower voltage, making it difficult to measure resistance changes accurately.

Gauge Factor

The gauge factor GF is defined as:

$$GF = \frac{\Delta R / R_G}{\epsilon}$$

where,

ΔR is the change in resistance caused by strain,

R_G is the resistance of the undeformed gauge, and

ϵ is strain.

For common metallic foil gauges, the gauge factor is usually a little over 2. For a single active gauge and three dummy resistors of the same resistance about the active gauge in a balanced Wheatstone bridge configuration, the output sensor voltage SV from the bridge is approximately:

$$SV = EV\frac{GF \cdot \epsilon}{4}$$

where,

EV is the bridge excitation voltage.

Foil gauges typically have active areas of about 2–10 mm^2 in size. With careful installation, the correct gauge, and the correct adhesive, strains up to at least 10% can be measured.

In Practice

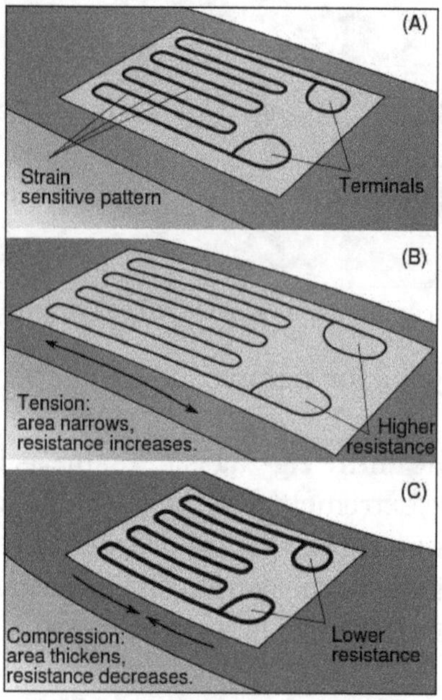

Visualization of the working concept behind the strain gauge on a beam under exaggerated bending.

An excitation voltage is applied to input leads of the gauge network, and a voltage reading is taken from the output leads. Typical input voltages are 5 V or 12 V and typical output readings are in millivolts.

Foil strain gauges are used in many situations. Different applications place different requirements on the gauge. In most cases the orientation of the strain gauge is significant.

Gauges attached to a load cell would normally be expected to remain stable over a period of years, if not decades; while those used to measure response in a dynamic experiment may only need to remain attached to the object for a few days, be energized for less than an hour, and operate for less than a second.

Strain gauges are attached to the substrate with a special glue. The type of glue depends on the required lifetime of the measurement system. For short term measurements (up to some weeks) cyanoacrylate glue is appropriate, for long lasting installation epoxy glue is required. Usually epoxy glue requires high temperature curing (at about 80-100 °C). The preparation of the surface where the strain gauge is to be glued is of the utmost importance. The surface must be smoothed (e.g. with very fine sand paper), deoiled with solvents, the solvent traces must then be removed and the strain gauge must be glued immediately after this to avoid oxidation or pollution of the prepared area. If these steps are not followed the strain gauge binding to the surface may be unreliable and unpredictable measurement errors may be generated.

Strain gauge based technology is utilized commonly in the manufacture of pressure sensors. The gauges used in pressure sensors themselves are commonly made from silicon, polysilicon, metal film, thick film, and bonded foil.

Variations in Temperature

Variations in temperature will cause a multitude of effects. The object will change in size by thermal expansion, which will be detected as a strain by the gauge. Resistance of the gauge will change, and resistance of the connecting wires will change.

Most strain gauges are made from a constantan alloy. Various constantan alloys and Karma alloys have been designed so that the temperature effects on the resistance of the strain gauge itself largely cancel out the resistance change of the gauge due to the thermal expansion of the object under test. Because different materials have different amounts of thermal expansion, self-temperature compensation (STC) requires selecting a particular alloy matched to the material of the object under test.

Strain gauges that are not self-temperature-compensated (such as isoelastic alloy) can be temperature compensated by use of the dummy gauge technique. A dummy gauge (identical to the active strain gauge) is installed on an unstrained sample of the same material as the test specimen. The sample with the dummy gauge is placed in thermal contact with the test specimen, adjacent to the active gauge. The dummy gauge is wired into a Wheatstone bridge on an adjacent arm to the active gauge so that the temperature effects on the active and dummy gauges cancel each other. (Murphy's law was originally coined in response to a set of gauges being incorrectly wired into a Wheatstone bridge).

Every material reacts when it heats up or when it cools down. This will cause strain gauges to register a deformation in the material which will make it change signal. To prevent this from happening strain gauges are made so they will compensate this change due to temperature. Dependent on the material of the surface where the strain gauge is assembled on, a different expansion can be measured.

Temperature effects on the lead wires can be cancelled by using a "3-wire bridge" or a "4-wire ohm circuit" (also called a "4-wire Kelvin connection").

In any case it is a good engineering practice to keep the Wheatstone bridge voltage drive low enough to avoid the self heating of the strain gauge. The self heating of the strain gauge depends on its mechanical characteristic (large strain gauges are less prone to self heating). Low voltage drive levels of the bridge reduce the sensitivity of the overall system.

Errors and Compensations

- Zero Offset: If the impedance of the four gauge arms are not exactly the same after bonding the gauge to the force collector, there will be a zero offset which can be compensated by introducing a parallel resistor to one or more of the gauge arms.
- Temperature coefficient of gauge factor (TCGF) is the change of sensitivity of the device to strain with change in temperature. This is generally compensated for by the introduction of a fixed resistance in the input leg, whereby the effective supplied voltage will decrease with a temperature increase, compensating for the increase in sensitivity with the temperature increase. This is known as modulus compensation in transducer circuits. As the temperature rises the load cell element becomes more elastic and therefore under a constant load will deform more and lead to an increase in output; but the load is still the same. The clever bit in all this is that the resistor in the bridge supply must be a temperature sensitive resistor that is matched to both the material to which the gauge is bonded and also to the gauge element material. The value of that resistor is dependent on both of those values and can be calculated. In simple terms if the output increases then the resistor value also increase thereby reducing the net voltage to the transducer. Get the resistor value right and you will see no change.
- Zero shift with temperature: If the TCGF of each gauge is not the same, there will be a zero shift with temperature. This is also caused by anomalies in the force collector. This is usually compensated for with one or more resistors strategically placed in the compensation network.
- Linearity is an error whereby the sensitivity changes across the pressure range. This is commonly a function of the force collection thickness selection for the intended pressure and the quality of the bonding.
- Hysteresis is an error of return to zero after pressure excursion.
- Repeatability: This error is sometimes tied-in with hysteresis but is across the pressure range.
- EMI induced errors: As strain gauges output voltage is in the mV range, even μV if the Wheatstone bridge voltage drive is kept low to avoid self heating of the element, special care must be taken in output signal amplification to avoid amplifying also the superimposed noise. A solution which is frequently adopted is to use "carrier frequency" amplifiers which convert the voltage variation into a frequency variation (as in VCOs) and have a narrow bandwidth thus reducing out of band EMI.

- Overloading: If a strain gauge is loaded beyond its design limit (measured in microstrain) its performance degrades and cannot be recovered. Normally good engineering practice suggests not to stress strain gauges beyond ±3000 microstrain.
- Humidity: If the wires connecting the strain gauge to the signal conditioner are not protected against humidity, such as bare wire, corrosion can occur, leading to parasitic resistance. This can allow currents to flow between the wires and the substrate to which the strain gauge is glued, or between the two wires directly, introducing an error which competes with the current flowing through the strain gauge. For this reason, high-current, low-resistance strain gauges (120 ohm) are less prone to this type of error. To avoid this error it is sufficient to protect the strain gauges wires with insulating enamel (e.g., epoxy or polyurethane type). Strain gauges with unprotected wires may be used only in a dry laboratory environment but not in an industrial one.

In some applications, strain gauges add mass and damping to the vibration profiles of the hardware they are intended to measure. In the turbomachinery industry, one used alternative to strain gauge technology in the measurement of vibrations on rotating hardware is the non-intrusive stress measurement system, which allows measurement of blade vibrations without any blade or disc-mounted hardware.

Geometries of Strain Gages

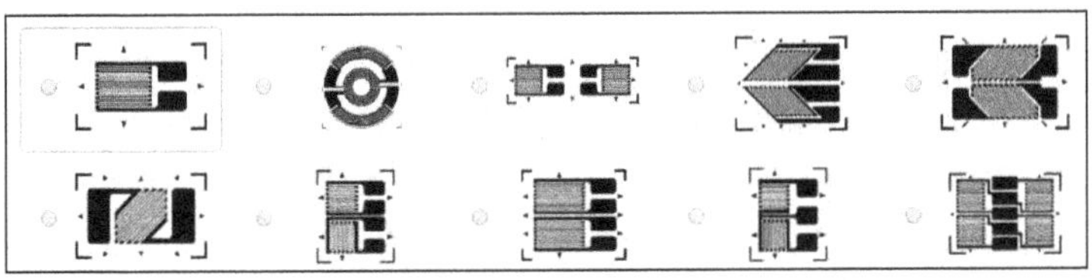

Geometries of strain gauges.

The following different kind of strain gages are available in the market:

- Linear strain gauges.
- Membrane Rosette strain gauges.
- Double linear strain gauges.
- Full bridge strain gauges.
- Shear strain gauges.
- Half bridge strain gauges.
- Column strain gauges.
- 45°-Rosette (3 measuring directions).
- 90°-Rosette (2 measuring directions).

Other Types

For measurements of small strain, semiconductor strain gauges, so called piezoresistors, are often preferred over foil gauges. A semiconductor gauge usually has a larger gauge factor than a foil gauge. Semiconductor gauges tend to be more expensive, more sensitive to temperature changes, and are more fragile than foil gauges.

Nanoparticle-based strain gauges emerge as a new promising technology. These resistive sensors whose active area is made by an assembly of conductive nanoparticles, such as gold or carbon, combine a high gauge factor, a large deformation range and a small electrical consumption due to their high impedance.

In biological measurements, especially blood flow and tissue swelling, a variant called mercury-in-rubber strain gauge is used. This kind of strain gauge consists of a small amount of liquid mercury enclosed in a small rubber tube, which is applied around e.g., a toe or leg. Swelling of the body part results in stretching of the tube, making it both longer and thinner, which increases electrical resistance.

Fiber optic sensing can be employed to measure strain along an optical fiber. Measurements can be distributed along the fiber, or taken at predetermined points on the fiber. The 2010 America's Cup boats Alinghi 5 and USA-17 both employ embedded sensors of this type. Other optical measuring techniques can be used to measure strains like electronic speckle pattern interferometry or digital image correlation.

Microscale strain gauges are widely used in microelectromechanical systems (MEMS) to measure strains such as those induced by force, acceleration, pressure or sound. As example, airbags in cars are often triggered with MEMS accelerometers. As alternative to piezo-resistant strain gauges, integrated optical ring resonators may be used to measure strain in microoptoelectromechanical systems (MOEMS). Capacitive strain gauges use a variable capacitor to indicate the level of mechanical deformation.

Vibrating wire strain gauges are used in geotechnical and civil engineering applications. The gauge consists of a vibrating, tensioned wire. The strain is calculated by measuring the resonant frequency of the wire (an increase in tension increases the resonant frequency).

Non-contact Strain Measurements

Strain can also be measured using digital image correlation (DIC). With this technique one or two cameras are used in conjunction with a DIC software to track features on the surface of components to detect small motion. The full strain map of the tested sample can be calculated, providing similar display as a finite-element analysis. This technique is used in many industries to replace traditional strain gauges or other sensors like extensometers, string pots, LVDT, accelerometers. The accuracy of commercially available DIC software typically ranges around 1/100 to 1/30 of a pixels for displacements measurements which result in strain sensitivity between 20 and 100 μm/m. The DIC technique allows to quickly measure shape, displacements and strain non-contact, avoiding some issues of traditional contacting methods, especially with impacts, high strain, high-temperature or high cycle fatigue testing.

Bourdon Tube

Bourdon Tubes are known for its very high range of differential pressure measurement in the range of almost 100,000 psi (700 MPa). It is an elastic type pressure transducer.

The device was invented by Eugene Bourdon in the year 1849. The basic idea behind the device is that, cross-sectional tubing when deformed in any way will tend to regain its circular form under the action of pressure. The bourdon pressure gauges used today have a slight elliptical cross-section and the tube is generally bent into a C-shape or arc length of about 27 degrees.

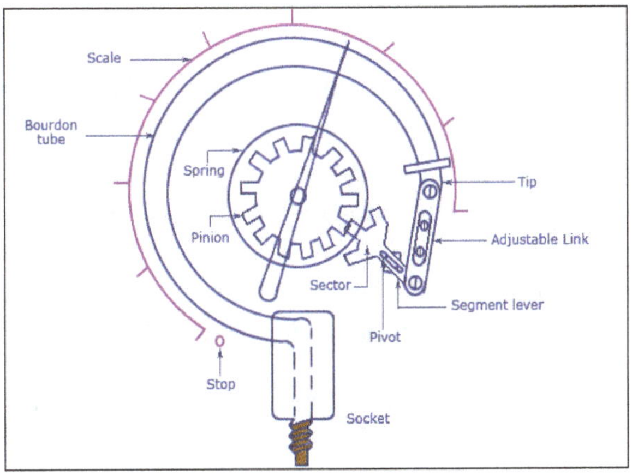

Bourdon Tube Pressure Gauge.

As seen in the figure, the pressure input is given to a socket which is soldered to the tube at the base. The other end or free end of the device is sealed by a tip. This tip is connected to a segmental lever through an adjustable length link. The lever length may also be adjustable. The segmental lever is suitably pivoted and the spindle holds the pointer as shown in the figure. A hair spring is sometimes used to fasten the spindle of the frame of the instrument to provide necessary tension for proper meshing of the gear teeth and thereby freeing the system from the backlash. Any error due to friction in the spindle bearings is known as lost motion. The mechanical construction has to be highly accurate in the case of a Bourdon Tube Gauge. If we consider a cross-section of the tube, its outer edge will have a larger surface than the inner portion. The tube walls will have a thickness between 0.01 and 0.05 inches.

Working

As the fluid pressure enters the bourdon tube, it tries to be reformed and because of a free tip available, this action causes the tip to travel in free space and the tube unwinds. The simultaneous actions of bending and tension due to the internal pressure make a non-linear movement of the free tip. This travel is suitable guided and amplified for the measurement of the internal pressure. But the main requirement of the device is that whenever the same pressure is applied, the movement of the tip should be the same and on withdrawal of the pressure the tip should return to the initial point.

A lot of compound stresses originate in the tube as soon as the pressure is applied. This makes the travel of the tip to be non-linear in nature. If the tip travel is considerably small, the stresses can be considered to produce a linear motion that is parallel to the axis of the link. The small linear tip movement is matched with a rotational pointer movement. This is known as multiplication, which can be adjusted by adjusting the length of the lever. For the same amount of tip travel, a shorter lever gives larger rotation. The approximately linear motion of the tip when converted to a circular motion with the link-lever and pinion attachment, a one-to-one correspondence between them may not occur and distortion results. This is known as angularity which can be minimized by adjusting the length of the link.

Other than C-type, bourdon gauges can also be constructed in the form of a helix or a spiral. The types are varied for specific uses and space accommodations, for better linearity and larger sensitivity. For thorough repeatability, the bourdon tubes materials must have good elastic or spring characteristics. The surrounding in which the process is carried out is also important as corrosive atmosphere or fluid would require a material which is corrosion proof. The commonly used materials are phosphor-bronze, silicon-bronze, beryllium-copper, inconel, and other C-Cr-Ni-Mo alloys, and so on.

In the case of forming processes, empirical relations are known to choose the tube size, shape and thickness and the radius of the C-tube. Because of the internal pressure, the near elliptic or rather the flattened section of the tube tries to expand as shown by the dotted line in the figure below (a) The same expansion lengthwise is shown in figure, (b) The arrangement of the tube, however forces an expansion on the outer surface and a compression on the inner surface, thus allowing the tube to unwind. This is shown in figure (c).

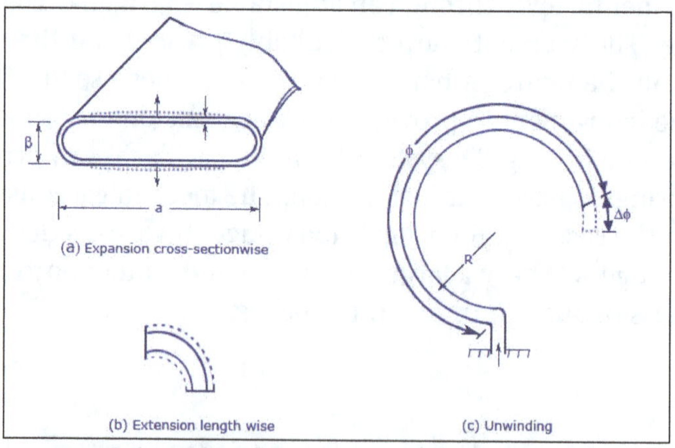

Expansion of Bourdon Tube due to internal pressure.

Like all elastic elements a bourdon tube also has some hysteresis in a given pressure cycle. By proper choice of material and its heat treatment, this may be kept to within

0.1 and 0.5 percent of the maximum pressure cycle. Sensitivity of the tip movement of a bourdon element without restraint can be as high as 0.01 percent of full range pressure reducing to 0.1 percent with restraint at the central pivot.

Temperature Transducer

Resistance Thermometer

Resistance thermometers, also called resistance temperature detectors (RTDs), are sensors used to measure temperature. Many RTD elements consist of a length of fine wire wrapped around a ceramic or glass core but other constructions are also used. The RTD wire is a pure material, typically platinum, nickel, or copper. The material has an accurate resistance/temperature relationship which is used to provide an indication of temperature. As RTD elements are fragile, they are often housed in protective probes.

RTDs, which have higher accuracy and repeatability, are slowly replacing thermocouples in industrial applications below 600 °C.

Resistance/Temperature Relationship of Metals

Common RTD sensing elements constructed of platinum, copper or nickel have a repeatable resistance versus temperature relationship (*R* vs *T*) and operating temperature range. The *R* vs *T* relationship is defined as the amount of resistance change of the sensor per degree of temperature change. The relative change in resistance (temperature coefficient of resistance) varies only slightly over the useful range of the sensor.

Platinum was proposed by Sir William Siemens as an element for a resistance temperature detector at the Bakerian lecture in 1871: it is a noble metal and has the most stable resistance–temperature relationship over the largest temperature range. Nickel elements have a limited temperature range because the amount of change in resistance per degree of change in temperature becomes very non-linear at temperatures over 300 °C (572 °F). Copper has a very linear resistance–temperature relationship; however, copper oxidizes at moderate temperatures and cannot be used over 150 °C (302 °F).

The significant characteristic of metals used as resistive elements is the linear approximation of the resistance versus temperature relationship between 0 and 100 °C. This temperature coefficient of resistance is denoted by α and is usually given in units of Ω/(Ω·°C):

$$\alpha = \frac{R_{100} - R_0}{100\,^\circ\mathrm{C} \cdot R_0},$$

where

R_0 is the resistance of the sensor at 0 °C,

R_{100} is the resistance of the sensor at 100 °C.

Pure platinum has α = 0.003925 Ω/(Ω·°C) in the 0 to 100 °C range and is used in the construction of laboratory-grade RTDs. Conversely, two widely recognized standards for industrial RTDs IEC 60751 and ASTM E-1137 specify α = 0.00385 Ω/(Ω·°C). Before these standards were widely adopted, several different α values were used. It is still possible to find older probes that are made with platinum that have α = 0.003916 Ω/(Ω·°C) and 0.003902 Ω/(Ω·°C).

These different α values for platinum are achieved by doping; basically, carefully introducing impurities into the platinum. The impurities introduced during doping become embedded in the lattice structure of the platinum and result in a different *R* vs. *T* curve and hence α value.

Calibration

To characterize the *R* vs *T* relationship of any RTD over a temperature range that represents the planned range of use, calibration must be performed at temperatures other than 0 °C and 100 °C. This is necessary to meet calibration requirements. Although RTDs are considered to be linear in operation, it must be proven that they are accurate with regard to the temperatures with which they will actually be used. Two common calibration methods are the fixed-point method and the comparison method.

- Fixed point calibration is used for the highest-accuracy calibrations by national metrology laboratories. It uses the triple point, freezing point or melting point of pure substances such as water, zinc, tin, and argon to generate a known and repeatable temperature. These cells allow the user to reproduce actual conditions of the ITS-90 temperature scale. Fixed-point calibrations provide extremely accurate calibrations (within ±0.001 °C). A common fixed-point calibration method for industrial-grade probes is the ice bath. The equipment is inexpensive, easy to use, and can accommodate several sensors at once. The ice point is designated as a secondary standard because its accuracy is ±0.005 °C (±0.009 °F), compared to ±0.001 °C (±0.0018 °F) for primary fixed points.

- Comparison calibrations is commonly used with secondary SPRTs and industrial RTDs. The thermometers being calibrated are compared to calibrated thermometers by means of a bath whose temperature is uniformly stable. Unlike fixed-point calibrations, comparisons can be made at any temperature between −100 °C and 500 °C (−148 °F to 932 °F). This method might be more cost-effective, since several sensors can be calibrated simultaneously with automated equipment. These electrically heated and well-stirred baths use silicone oils and molten salts as the medium for the various calibration temperatures.

Element Types

The three main categories of RTD sensors are thin-film, wire-wound, and coiled elements. While these types are the ones most widely used in industry, other more exotic shapes are used; for example, carbon resistors are used at ultra-low temperatures (−273 °C to −173 °C).

- Carbon resistor elements are cheap and widely used. They have very reproducible results at low temperatures. They are the most reliable form at extremely low temperatures. They generally do not suffer from significant hysteresis or strain gauge effects.

- Strain-free elements use a wire coil minimally supported within a sealed housing filled with an inert gas. These sensors work up to 961.78 °C and are used in the SPRTs that define ITS-90. They consist of platinum wire loosely coiled over a support structure, so the element is free to expand and contract with temperature. They are very susceptible to shock and vibration, as the loops of platinum can sway back and forth, causing deformation.

- Thin-film elements have a sensing element that is formed by depositing a very thin layer of resistive material, normally platinum, on a ceramic substrate (plating). This layer is usually just 10 to 100 ångströms (1 to 10 nanometers) thick. This film is then coated with an epoxy or glass that helps protect the deposited film and also acts as a strain relief for the external lead wires. Disadvantages of this type are that they are not as stable as their wire-wound or coiled counterparts. They also can only be used over a limited temperature range due to the different expansion rates of the substrate and resistive deposited giving a "strain gauge" effect that can be seen in the resistive temperature coefficient. These elements work with temperatures to 300 °C (572 °F) without further packaging, but can operate up to 600 °C (1,112 °F) when suitably encapsulated in glass or ceramic. Special high-temperature RTD elements can be used up to 900 °C (1,652 °F) with the right encapsulation.

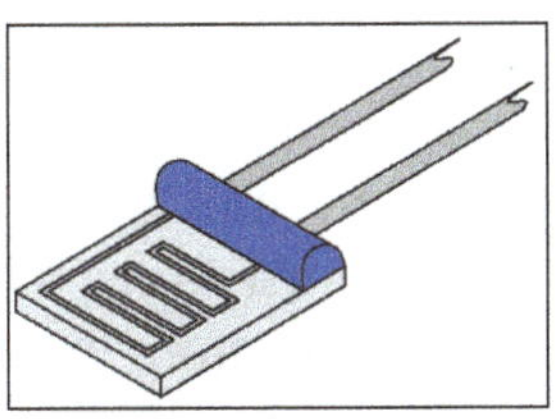

Thin-film PRT.

- Wire-wound elements can have greater accuracy, especially for wide temperature ranges. The coil diameter provides a compromise between mechanical stability and allowing expansion of the wire to minimize strain and consequential drift. The sensing wire is wrapped around an insulating mandrel or core. The winding core can be round or flat, but must be an electrical insulator. The coefficient of thermal expansion of the winding core material is matched to the sensing wire to minimize any mechanical strain. This strain on the element wire will result in a thermal measurement error. The sensing wire is connected to a larger wire, usually referred to as the element lead or wire. This wire is selected to be compatible with the sensing wire, so that the combination does not generate an emf that would distort the thermal measurement. These elements work with temperatures to 660 °C.

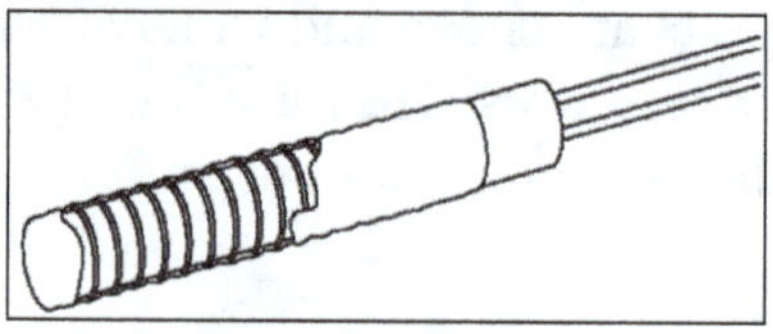

Wire-wound PRT.

- Coiled elements have largely replaced wire-wound elements in industry. This design has a wire coil that can expand freely over temperature, held in place by some mechanical support, which lets the coil keep its shape. This "strain free" design allows the sensing wire to expand and contract free of influence from other materials; in this respect it is similar to the SPRT, the primary standard upon which ITS-90 is based, while providing the durability necessary for industrial use. The basis of the sensing element is a small coil of platinum sensing wire. This coil resembles a filament in an incandescent light bulb. The housing or mandrel is a hard fired ceramic oxide tube with equally spaced bores that run transverse to the axes. The coil is inserted in the bores of the mandrel and then packed with a very finely ground ceramic powder. This permits the sensing wire to move, while still remaining in good thermal contact with the process. These elements work with temperatures to 850 °C.

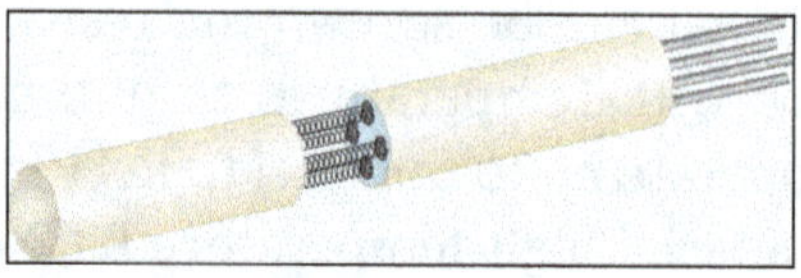

Coil-element PRT.

The current international standard that specifies tolerance and the temperature-to-electrical resistance relationship for platinum resistance thermometers (PRTs) is IEC 60751:2008; ASTM E1137 is also used in the United States. By far the most common devices used in industry have a nominal resistance of 100 ohms at 0 °C and are called Pt100 sensors ("Pt" is the symbol for platinum, "100" for the resistance in ohms at 0 °C). It is also possible to get Pt1000 sensors, where 1000 is for the resistance in ohms at 0 °C. The sensitivity of a standard 100 Ω sensor is a nominal 0.385 Ω/°C. RTDs with a sensitivity of 0.375 and 0.392 Ω/°C, as well as a variety of others, are also available.

Function

Resistance thermometers are constructed in a number of forms and offer greater stability, accuracy and repeatability in some cases than thermocouples. While thermocouples use the Seebeck effect to generate a voltage, resistance thermometers use electrical resistance and require a power source to operate. The resistance ideally varies nearly linearly with temperature per the Callendar–Van Dusen equation.

The platinum detecting wire needs to be kept free of contamination to remain stable. A platinum wire or film is supported on a former in such a way that it gets minimal differential expansion or other strains from its former, yet is reasonably resistant to vibration. RTD assemblies made from iron or copper are also used in some applications. Commercial platinum grades exhibit a temperature coefficient of resistance 0.00385/°C (0.385%/°C) (European Fundamental Interval). The sensor is usually made to have a resistance of 100 Ω at 0 °C. This is defined in BS EN 60751:1996 (taken from IEC 60751:1995). The American Fundamental Interval is 0.00392/°C, based on using a purer grade of platinum than the European standard. The American standard is from the Scientific Apparatus Manufacturers Association (SAMA), who are no longer in this standards field. As a result, the "American standard" is hardly the standard even in the US.

Lead-wire resistance can also be a factor; adopting three- and four-wire, instead of two-wire, connections can eliminate connection-lead resistance effects from measurements; three-wire connection is sufficient for most purposes and is an almost universal industrial practice. Four-wire connections are used for the most precise applications.

Advantages and Disadvantages

The advantages of platinum resistance thermometers include:

- High accuracy.
- Low drift.
- Wide operating range.
- Suitability for precision applications.

Disadvantages

RTDs in industrial applications are rarely used above 660 °C. At temperatures above 660 °C it becomes increasingly difficult to prevent the platinum from becoming contaminated by impurities from the metal sheath of the thermometer. This is why laboratory standard thermometers replace the metal sheath with a glass construction. At very low temperatures, say below −270 °C (3 K), because there are very few phonons, the resistance of an RTD is mainly determined by impurities and boundary scattering and thus basically independent of temperature. As a result, the sensitivity of the RTD is essentially zero and therefore not useful.

Compared to thermistors, platinum RTDs are less sensitive to small temperature changes and have a slower response time. However, thermistors have a smaller temperature range and stability.

RTDs vs. Thermocouples

The two most common ways of measuring temperatures for industrial applications are with resistance temperature detectors (RTDs) and thermocouples. Choice between them is usually determined by four factors:

- Temperature: If process temperatures are between −200 and 500 °C (−328.0 and 932.0 °F), an industrial RTD is the preferred option. Thermocouples have a range of −180 to 2,320 °C (−292.0 to 4,208.0 °F), so for temperatures above 500 °C (932 °F) they are the only contact temperature measurement device.
- Response time: If the process requires a very fast response to temperature changes (fractions of a second as opposed to seconds), then a thermocouple is the best choice. Time response is measured by immersing the sensor in water moving at 1 m/s (3.3 ft/s) with a 63.2% step change.
- Size: A standard RTD sheath is 3.175 to 6.35 mm (0.1250 to 0.2500 in) in diameter; sheath diameters for thermocouples can be less than 1.6 mm (0.063 in).
- Accuracy and stability requirements: If a tolerance of 2 °C is acceptable and the highest level of repeatability is not required, a thermocouple will serve. RTDs are capable of higher accuracy and can maintain stability for many years, while thermocouples can drift within the first few hours of use.

Construction

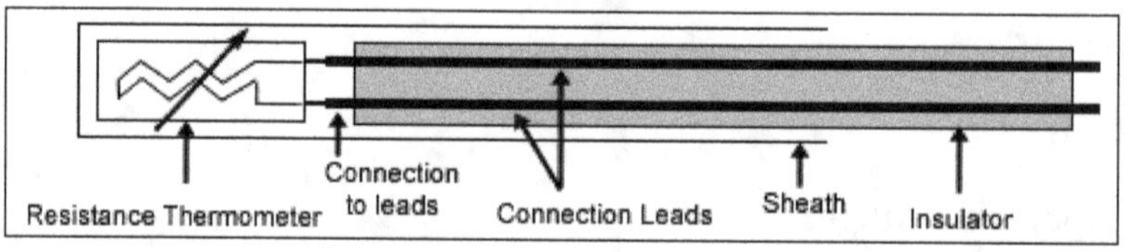

These elements nearly always require insulated leads attached. PVC, silicone rubber or PTFE insulators are used at temperatures below about 250 °C. Above this, glass fibre or ceramic are used. The measuring point, and usually most of the leads, require a housing or protective sleeve, often made of a metal alloy that is chemically inert to the process being monitored. Selecting and designing protection sheaths can require more care than the actual sensor, as the sheath must withstand chemical or physical attack and provide convenient attachment points.

Wiring Configurations

Two-wire Configuration

The simplest resistance-thermometer configuration uses two wires. It is only used when high accuracy is not required, as the resistance of the connecting wires is added to that of the sensor, leading to errors of measurement. This configuration allows use of 100 meters of cable. This applies equally to balanced bridge and fixed bridge system.

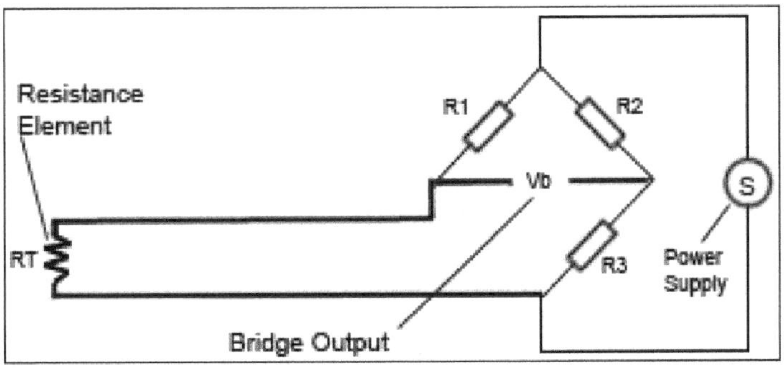

For a balanced bridge usual setting is with R2 = R1, and R3 around the middle of the range of the RTD. So for example, if we are going to measure between 0 and 100 °C (32 and 212 °F), RTD resistance will range from 100 Ω to 138.5 Ω. We would choose R1 = 120 Ω. In that way we get a small measured voltage in the bridge.

Three-wire Configuration

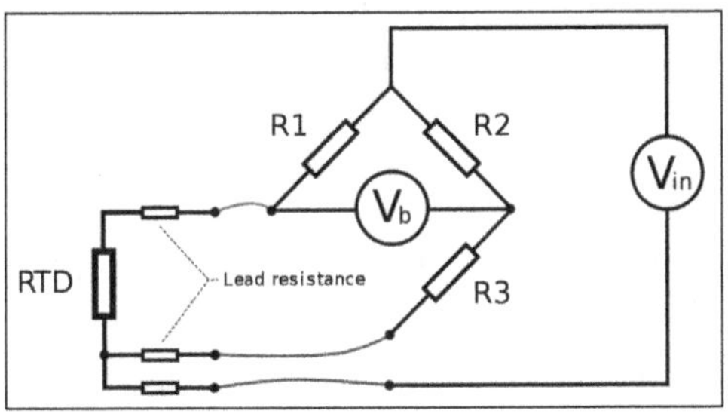

In order to minimize the effects of the lead resistances, a three-wire configuration can be used. The suggested setting for the configuration shown, is with R1 = R2, and R3 around the middle of the range of the RTD. Looking at the Wheatstone bridge circuit shown, the voltage drop on the lower left hand side is V_rtd + V_lead, and on the lower right hand size is V_R3 + V_lead, therefore the bridge voltage (V_b) is the difference, V_rtd - V_R3. The voltage drop due to the lead resistance has been cancelled out. This always applies if R1=R2, and R1, R2 >> RTD, R3. R1 and R2 can serve the use of limiting the current through the RTD, for example for a PT100, limiting to 1mA, and 5V, would suggest a limiting resistance of approximately R1 = R2 = 5/0.001 = 5,000 Ohms.

Four-wire Configuration

The four-wire resistance configuration increases the accuracy of measurement of resistance. Four-terminal sensing eliminates voltage drop in the measuring leads as a contribution to error. To increase accuracy further, any residual thermoelectric voltages generated by different wire types or screwed connections are eliminated by reversal of the direction of the 1 mA current and the leads to the DVM (digital voltmeter). The thermoelectric voltages will be produced in one direction only. By averaging the reversed measurements, the thermoelectric error voltages are cancelled out.

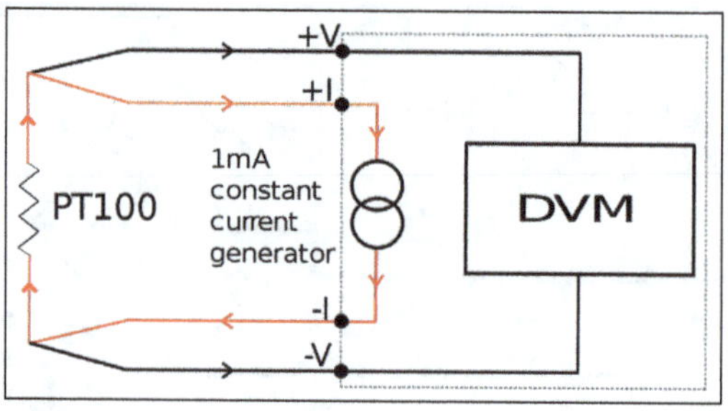

Classifications of RTDs

The highest-accuracy of all PRTs are the Ultra Precise Platinum Resistance Thermometers (UPRTs). This accuracy is achieved at the expense of durability and cost. The UPRT elements are wound from reference-grade platinum wire. Internal lead wires are usually made from platinum, while internal supports are made from quartz or fused silica. The sheaths are usually made from quartz or sometimes Inconel, depending on temperature range. Larger-diameter platinum wire is used, which drives up the cost and results in a lower resistance for the probe (typically 25.5 Ω). UPRTs have a wide temperature range (−200 °C to 1000 °C) and are approximately accurate to ±0.001 °C over the temperature range. UPRTs are only appropriate for laboratory use.

Another classification of laboratory PRTs is Standard Platinum Resistance Thermometers (Standard SPRTs). They are constructed like the UPRT, but the materials are more cost-effective. SPRTs commonly use reference-grade, high-purity smaller-diameter platinum wire, metal sheaths and ceramic type insulators. Internal lead wires are usually a nickel-based alloy. Standard PRTs are more limited in temperature range (−200 °C to 500 °C) and are approximately accurate to ±0.03 °C over the temperature range.

Industrial PRTs are designed to withstand industrial environments. They can be almost as durable as a thermocouple. Depending on the application, industrial PRTs can use thin-film or coil-wound elements. The internal lead wires can range from PTFE-insulated stranded nickel-plated copper to silver wire, depending on the sensor size and application. Sheath material is typically stainless steel; higher-temperature applications may demand Inconel. Other materials are used for specialized applications.

Standard Resistance Thermometer Data

Temperature sensors are usually supplied with thin-film elements. The resistance elements are rated in accordance with BS EN 60751:2008 as:

Tolerance class	Valid range
F 0.3	−50 to +500 °C
F 0.15	−30 to +300 °C
F 0.1	0 to +150 °C

Resistance-thermometer elements functioning up to 1000 °C can be supplied. The relation between temperature and resistance is given by the Callendar-Van Dusen equation:

$$R_T = R_0\left[1 + AT + BT^2 + CT^3(T - 100)\right](-200\ ^\circ C < T < 0\ ^\circ C),$$

$$R_T = R_0\left[1 + AT + BT^2\right](0\ ^\circ C \le T < 850\ ^\circ C).$$

Here R_T is the resistance at temperature *T*, R_0 is the resistance at 0 °C, and the constants (for an $\alpha = 0.00385$ platinum RTD) are:

$$A = 3.9083 \times 10^{-3}\ ^\circ C^{-1},$$

$$B = -5.775 \times 10^{-7}\ ^\circ C^{-2},$$

$$C = -4.183 \times 10^{-12}\ ^\circ C^{-4}.$$

Since the *B* and *C* coefficients are relatively small, the resistance changes almost linearly with the temperature.

For positive temperature, solution of the quadratic equation yields the following relationship between temperature and resistance:

$$T = \frac{-A + \sqrt{A^2 - 4B\left(1 - \frac{R_T}{R_0}\right)}}{2B}.$$

Then for a four-wire configuration with a 1 mA precision current source the relationship between temperature and measured voltage V_T is,

$$T = \frac{-A + \sqrt{A^2 - 40B(0.1 - V_T)}}{2B}.$$

Thermistor

A thermistor is a type of resistor whose resistance is dependent on temperature, more so than in standard resistors. The word is a combination of thermal and resistor. Thermistors are widely used as inrush current limiters, temperature sensors (negative temperature coefficient or NTC type typically), self-resetting overcurrent protectors, and self-regulating heating elements (positive temperature coefficient or PTC type typically).

Thermistors are of two opposite fundamental types:

- With NTC thermistors, resistance decreases as temperature rises. An NTC is commonly used as a temperature sensor, or in series with a circuit as an inrush current limiter.

- With PTC thermistors, resistance increases as temperature rises. PTC thermistors are commonly installed in series with a circuit, and used to protect against overcurrent conditions, as resettable fuses.

Thermistors are generally produced using powdered metal oxides. With vastly improved formulas and techniques over the past 20 years, NTC thermistors can now achieve precision accuracies over wide temperature ranges such as ±0.1 °C or ±0.2 °C from 0 °C to 70 °C with excellent long-term stability. NTC thermistor elements come in many styles such as axial-leaded glass-encapsulated (DO-35, DO-34 and DO-41 diodes), glass-coated chips, epoxy-coated with bare or insulated lead wire and surface-mount, as well as rods and discs. The typical operating temperature range of a thermistor is −55 °C to +150 °C, though some glass-body thermistors have a maximal operating temperature of +300 °C.

Thermistors differ from resistance temperature detectors (RTDs) in that the material used in a thermistor is generally a ceramic or polymer, while RTDs use pure metals. The temperature response is also different; RTDs are useful over larger temperature ranges, while thermistors typically achieve a greater precision within a limited temperature range, typically −90 °C to 130 °C.

Basic Operation

Assuming, as a first-order approximation, that the relationship between resistance and temperature is linear, then

$$\Delta R = k\Delta T,$$

where,

- ΔR, change in resistance,
- ΔT, change in temperature,
- k, first-order temperature coefficient of resistance.

Thermistors can be classified into two types, depending on the sign of k. If k is positive, the resistance increases with increasing temperature, and the device is called a positive temperature coefficient (PTC) thermistor, or posistor. If k is negative, the resistance decreases with increasing temperature, and the device is called a negative temperature coefficient (NTC) thermistor. Resistors that are not thermistors are designed to have a k as close to 0 as possible, so that their resistance remains nearly constant over a wide temperature range.

Instead of the temperature coefficient k, sometimes the temperature coefficient of resistance α_T ("alpha sub T") is used. It is defined as,

$$\alpha_T = \frac{1}{R(T)}\frac{dR}{dT}.$$

This α_T coefficient should not be confused with the a parameter below.

Steinhart–Hart equation

In practical devices, the linear approximation model is accurate only over a limited temperature range. Over wider temperature ranges, a more complex resistance–temperature transfer function provides a more faithful characterization of the performance. The Steinhart–Hart equation is a widely used third-order approximation:

$$\frac{1}{T} = a + b\ln R + c(\ln R)^3,$$

where a, b and c are called the Steinhart–Hart parameters and must be specified for each device. T is the absolute temperature, and R is the resistance. To give resistance as a function of temperature, the above can be rearranged into,

$$R = \exp\left[(x-y)^{1/3} - (x+y)^{1/3}\right],$$

where

$$y = \frac{1}{2c}\left(a - \frac{1}{T}\right),$$

$$x = \sqrt{\left(\frac{b}{3c}\right)^3 + y^2}.$$

The error in the Steinhart–Hart equation is generally less than 0.02 °C in the measurement of temperature over a 200 °C range. As an example, typical values for a thermistor with a resistance of 3 kΩ at room temperature (25 °C = 298.15 K) are:

$$a = 1.40 \times 10^{-3},$$

$$b = 2.37 \times 10^{-4},$$

$$c = 9.90 \times 10^{-8}.$$

B or *β* Parameter Equation

NTC thermistors can also be characterised with the B (or β) parameter equation, which is essentially the Steinhart–Hart equation with $a = 1/T_0 - (1/B)\ln R_0$, $b = 1/B$ and $c = 0$,

$$\frac{1}{T} = \frac{1}{T_0} + \frac{1}{B}\ln\frac{R}{R_0},$$

where the temperatures are in kelvins, and R_0 is the resistance at temperature T_0 (25 °C = 298.15 K). Solving for R yields,

$$R = R_0 e^{B\left(\frac{1}{T} - \frac{1}{T_0}\right)}$$

or, alternatively,

$$R = r_\infty e^{B/T},$$

where $r_\infty = R_0 e^{-B/T_0}$.

This can be solved for the temperature:

$$T = \frac{B}{\ln(R / r_\infty)}.$$

The *B*-parameter equation can also be written as $\ln R = B / T + \ln r_\infty$. This can be used to convert the function of resistance vs. temperature of a thermistor into a linear function of $\ln R$ vs. $1/T$. The average slope of this function will then yield an estimate of the value of the *B* parameter.

Conduction Model

NTC (Negative Temperature Coefficient)

A failed (blown) NTC thermistor that worked as an inrush current limiter in a switched-mode power supply.

Many NTC thermistors are made from a pressed disc, rod, plate, bead or cast chip of semiconducting material such as sintered metal oxides. They work because raising the temperature of a semiconductor increases the number of active charge carriers it promotes them into the conduction band. The more charge carriers that are available, the more current a material can conduct. In certain materials like ferric oxide (Fe_2O_3) with titanium (Ti) doping an n-type semiconductor is formed and the charge carriers are electrons. In materials such as nickel oxide (NiO) with lithium (Li) doping a p-type semiconductor is created, where holes are the charge carriers.

This is described in the formula,

$$I = n \cdot A \cdot v \cdot e,$$

where

- I = electric current (amperes).
- n = density of charge carriers (count/m³).

- A = cross-sectional area of the material (m^2).
- v = drift velocity of electrons (m/s).
- e = charge of an electron ($e = 1.602 \times 10^{-19}$ coulomb).

Over large changes in temperature, calibration is necessary. Over small changes in temperature, if the right semiconductor is used, the resistance of the material is linearly proportional to the temperature. There are many different semiconducting thermistors with a range from about 0.01 kelvin to 2,000 kelvins (−273.14 °C to 1,700 °C).

The IEC standard symbol for a NTC thermistor includes a "−t°" under the rectangle.

PTC (Positive Temperature Coefficient)

Most PTC thermistors are made from doped polycrystalline ceramic (containing barium titanate ($BaTiO_3$) and other compounds) which have the property that their resistance rises suddenly at a certain critical temperature. Barium titanate is ferroelectric and its dielectric constant varies with temperature. Below the Curie point temperature, the high dielectric constant prevents the formation of potential barriers between the crystal grains, leading to a low resistance. In this region the device has a small negative temperature coefficient. At the Curie point temperature, the dielectric constant drops sufficiently to allow the formation of potential barriers at the grain boundaries, and the resistance increases sharply with temperature. At even higher temperatures, the material reverts to NTC behaviour.

Another type of thermistor is a silistor, a thermally sensitive silicon resistor. Silistors employ silicon as the semiconductive component material. Unlike ceramic PTC thermistors, silistors have an almost linear resistance-temperature characteristic.

Barium titanate thermistors can be used as self-controlled heaters; for a given voltage, the ceramic will heat to a certain temperature, but the power used will depend on the heat loss from the ceramic.

The dynamics of PTC thermistors being powered also is extremely useful. When first connected to a voltage source, a large current corresponding to the low, cold, resistance flows, but as the thermistor self-heats, the current is reduced until a limiting current (and corresponding peak device temperature) is reached. The current-limiting effect can replace fuses. In the degaussing circuits of many CRT monitors and televisions an appropriately chosen thermistor is connected in series with the degaussing coil. This results in a smooth current decrease for an improved degaussing effect. Some of these degaussing circuits have auxiliary heating elements to heat the thermistor (and reduce the resulting current) further.

Another type of PTC thermistor is the polymer PTC, which is sold under brand names such as "Polyswitch" "Semifuse", and "Multifuse". This consists of plastic with carbon grains embedded in it. When the plastic is cool, the carbon grains are all in contact with each other, forming a conductive path through the device. When the plastic heats up, it expands, forcing the carbon grains apart, and causing the resistance of the device to rise, which then causes increased heating and rapid resistance increase. Like the $BaTiO_3$ thermistor, this device has a highly nonlinear resistance/temperature response useful for thermal or circuit control, not for temperature measurement. Besides circuit elements used to limit current, self-limiting heaters can be made in the form of wires or strips, useful for heat tracing. PTC thermistors 'latch' into a hot / high resistance state: once hot, they stay in that high resistance state, until cooled. The effect can be used as a primitive latch/memory circuit, the effect being enhanced by using two PTC thermistors in series, with one thermistor cool, and the other thermistor hot.

The IEC standard symbol for a PTC thermistor includes a "+t°" under the rectangle.

Self-heating Effects

When a current flows through a thermistor, it generates heat, which raises the temperature of the thermistor above that of its environment. If the thermistor is being used to measure the temperature of the environment, this electrical heating may introduce a significant error if a correction is not made. Alternatively, this effect itself can be exploited. It can, for example, make a sensitive air-flow device employed in a sailplane rate-of-climb instrument, the electronic variometer, or serve as a timer for a relay as was formerly done in telephone exchanges.

The electrical power input to the thermistor is just,

$$P_E = IV,$$

where I is current, and V is the voltage drop across the thermistor. This power is converted to heat, and this heat energy is transferred to the surrounding environment. The rate of transfer is well described by Newton's law of cooling:

$$P_T = K(T(R) - T_0),$$

where $T(R)$ is the temperature of the thermistor as a function of its resistance R, T_0 is the temperature of the surroundings, and K is the dissipation constant, usually expressed in units of milliwatts per degree Celsius. At equilibrium, the two rates must be equal:

$$P_E = P_T.$$

The current and voltage across the thermistor depend on the particular circuit configuration. As a simple example, if the voltage across the thermistor is held fixed, then by Ohm's law we have $I = V / R$, and the equilibrium equation can be solved for the ambient temperature as a function of the measured resistance of the thermistor:

$$T_0 = T(R) - \frac{V^2}{KR}.$$

The dissipation constant is a measure of the thermal connection of the thermistor to its surroundings. It is generally given for the thermistor in still air and in well-stirred oil. Typical values for a small glass-bead thermistor are 1.5 mW/°C in still air and 6.0 mW/°C in stirred oil. If the temperature of the environment is known beforehand, then a thermistor may be used to measure the value of the dissipation constant. For example, the thermistor may be used as a flow-rate sensor, since the dissipation constant increases with the rate of flow of a fluid past the thermistor.

The power dissipated in a thermistor is typically maintained at a very low level to ensure insignificant temperature measurement error due to self-heating. However, some thermistor applications depend upon significant "self-heating" to raise the body temperature of the thermistor well above the ambient temperature, so the sensor then detects even subtle changes in the thermal conductivity of the environment. Some of these applications include liquid-level detection, liquid-flow measurement and air-flow measurement.

Applications

PTC

- As current-limiting devices for circuit protection, as replacements for fuses. Current through the device causes a small amount of resistive heating. If the current is large enough to generate more heat than the device can lose to its surroundings, the device heats up, causing its resistance to increase. This creates a self-reinforcing effect that drives the resistance upwards, therefore limiting the current.
- As timers in the degaussing coil circuit of most CRT displays. When the display unit is initially switched on, current flows through the thermistor and degaussing coil. The coil and thermistor are intentionally sized so that the current flow will heat the thermistor to the point that the degaussing coil shuts off in under a second. For effective degaussing, it is necessary that the magnitude of the alternating magnetic field produced by the degaussing coil decreases smoothly and continuously, rather than sharply switching off or decreasing in steps; the PTC thermistor accomplishes this naturally as it heats up. A degaussing circuit using a PTC thermistor is simple, reliable (for its simplicity), and inexpensive.
- As heater in automotive industry to provide additional heat inside cabin with diesel engine or to heat diesel in cold climatic conditions before engine injection.
- In temperature compensated synthesizer voltage controlled oscillators.
- In lithium battery protection circuits.
- In an electrically actuated Wax motor to provide the heat necessary to expand the wax.
- Many electric motors and dry type power transformers incorporate PTC thermistors in their windings. When used in conjunction with a monitoring relay they

provide over temperature protection to prevent insulation damage. The equipment manufacturer selects a thermistor with a highly non-linear response curve where resistance increases dramatically at the maximum allowable winding temperature, causing the relay to operate.

NTC

- As a resistance thermometer for low-temperature measurements of the order of 10 K.
- As an inrush current limiter device in power supply circuits, they present a higher resistance initially, which prevents large currents from flowing at turn-on, and then heat up and become much lower resistance to allow higher current flow during normal operation. These thermistors are usually much larger than measuring type thermistors, and are purposely designed for this application.
- As sensors in automotive applications to monitor fluid temperatures like the engine coolant, cabin air, external air or engine oil temperature, and feed the relative readings to control units like the ECU and to the dashboard.
- To monitor the temperature of an incubator.
- Thermistors are also commonly used in modern digital thermostats and to monitor the temperature of battery packs while charging.
- Thermistors are often used in the hot ends of 3D printers; they monitor the heat produced and allow the printer's control circuitry to keep a constant temperature for melting the plastic filament.
- In the food handling and processing industry, especially for food storage systems and food preparation. Maintaining the correct temperature is critical to prevent foodborne illness.
- Throughout the consumer appliance industry for measuring temperature. Toasters, coffee makers, refrigerators, freezers, hair dryers, etc. all rely on thermistors for proper temperature control.
- NTC thermistors come in bare and lugged forms, the former is for point sensing to achieve high accuracy for specific points, such as laser diode die, etc.
- For measurement of temperature profile inside the sealed cavity of a convective (thermal) inertial sensor.
- Thermistor Probe Assemblies offer protection of the sensor in harsh environments. The thermistor sensing element can be packaged into a variety of enclosures for use in industries such as HVAC/R, Building Automation, Pool/Spa, Energy and Industrial Electronics. Enclosures can be made out of stainless steel, aluminum, copper brass or plastic and configurations include threaded

(NPT, etc), flanged (with mounting holes for ease of installation) and straight (flat tip, pointed tip, radius tip, etc.). Thermistor probe assemblies are very rugged and are highly customizable to fit the application needs. Probe assemblies have gained in popularity over the years as improvements in research, engineering and manufacturing techniques have been made.

Flow Transducers

Orifice Plate

An orifice plate is a device used for measuring flow rate, for reducing pressure or for restricting flow (in the latter two cases it is often called a restriction plate).

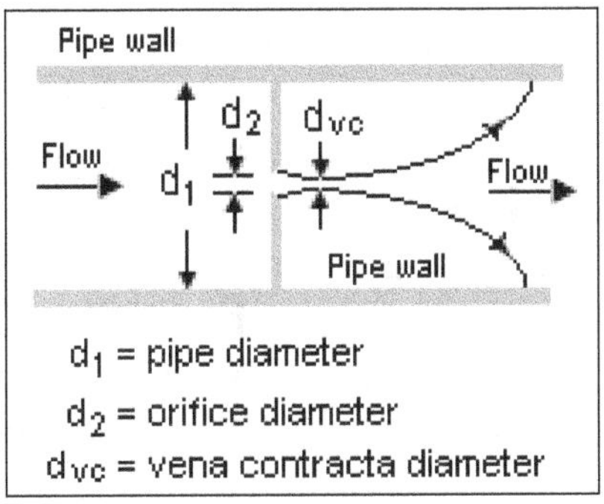

Orifice plate showing vena contracta.

An orifice plate is a thin plate with a hole in it, which is usually placed in a pipe. When a fluid (whether liquid or gaseous) passes through the orifice, its pressure builds up slightly upstream of the orifice but as the fluid is forced to converge to pass through the hole, the velocity increases and the fluid pressure decreases. A little downstream of the orifice the flow reaches its point of maximum convergence, the vena contracta where the velocity reaches its maximum and the pressure reaches its minimum. Beyond that, the flow expands, the velocity falls and the pressure increases. By measuring the difference in fluid pressure across tappings upstream and downstream of the plate, the flow rate can be obtained from Bernoulli's equation using coefficients established from extensive research.

In general, the mass flow rate q_m measured in kg/s across an orifice can be described as,

$$q_m = \frac{C_d}{\sqrt{1-\beta^4}} \epsilon \frac{\pi}{4} d^2 \sqrt{2\Delta p * \rho_1}$$

where:

- C_d = Coefficient of discharge, dimensionless, typically between 0.6 and 0.85, depending on the orifice geometry and tappings.

- β = Diameter ratio of orifice diameter d to pipe diameter D, dimensionless.
- ϵ = Expansibility factor, 1 for incompressible gases and most liquids, and decreasing with pressure ratio across the orifice, dimensionless.
- d = Internal orifice diameter under operating conditions, m.
- ρ_1 = Fluid density in plane of upstream tapping, kg/m³.
- Δp = Differential pressure measured across the orifice, Pa.

The overall pressure loss in the pipe due to an orifice plate is lower than the measured pressure, typically by a factor of $1-\beta^{1.9}$.

Application

Orifice plates are most commonly used to measure flow rates in pipes, when the fluid is single-phase (rather than being a mixture of gases and liquids, or of liquids and solids) and well-mixed, the flow is continuous rather than pulsating, the fluid occupies the entire pipe (precluding silt or trapped gas), the flow profile is even and well-developed and the fluid and flow rate meet certain other conditions. Under these circumstances and when the orifice plate is constructed and installed according to appropriate standards, the flow rate can easily be determined using published formulae based on substantial research and published in industry, national and international standards.

An orifice plate is called a calibrated orifice if it has been calibrated with an appropriate fluid flow and a traceable flow measurement device.

Plates are commonly made with sharp-edged circular orifices and installed concentric with the pipe and with pressure tappings at one of three standard pairs of distances upstream and downstream of the plate; these types are covered by ISO 5167 and other major standards. There are many other possibilities. The edges may be rounded or conical, the plate may have an orifice the same size as the pipe except for a segment at top or bottom which is obstructed, the orifice may be installed eccentric to the pipe, and the pressure tappings may be at other positions. Variations on these possibilities are covered in various standards and handbooks. Each combination gives rise to different coefficients of discharge which can be predicted so long as various conditions are met, conditions which differ from one type to another.

Once the orifice plate is designed and installed, the flow rate can often be indicated with an acceptably low uncertainty simply by taking the square root of the differential pressure across the orifice's pressure tappings and applying an appropriate constant.

Orifice plates are also used to reduce pressure or restrict flow, in which case they are often called restriction plates.

Pressure Tappings

There are three standard positions for pressure tappings (also called taps), commonly named as follows:

- Corner taps placed immediately upstream and downstream of the plate; convenient when the plate is provided with an orifice carrier incorporating tappings.
- D and D/2 taps *or* radius taps placed one pipe diameter upstream and half a pipe diameter downstream of the plate; these can be installed by welding bosses to the pipe.
- Flange taps placed 25.4 mm (1 inch) upstream and downstream of the plate, normally within specialised pipe flanges.

These types are covered by ISO 5167 and other major standards. Other types include:

- 2½D and 8D taps *or* recovery taps placed 2.5 pipe diameters upstream and 8 diameters downstream, at which point the measured differential is equal to the unrecoverable pressure loss caused by the orifice.
- Vena contracta tappings placed one pipe diameter upstream and at a position 0.3 to 0.9 diameters downstream, depending on the orifice type and size relative to the pipe, in the plane of minimum fluid pressure.

The measured differential pressure differs for each combination and so the coefficient of discharge used in flow calculations depends partly on the tapping positions.

The simplest installations use single tappings upstream and downstream, but in some circumstances these may be unreliable; they might be blocked by solids or gas-bubbles, or the flow profile might be uneven so that the pressures at the tappings are higher or lower than the average in those planes. In these situations multiple tappings can be used, arranged circumferentially around the pipe and joined by a piezometer ring, or (in the case of corner taps) annular slots running completely round the internal circumference of the orifice carrier.

Plate

Standards and handbooks are mainly concerned with sharp-edged thin plates. In these, the leading edge is sharp and free of burrs and the cylindrical section of the orifice is short, either because the entire plate is thin or because the downstream edge of the plate is bevelled. Exceptions include the quarter-circle *or* quadrant-edge orifice, which has a fully rounded leading edge and no cylindrical section, and the conical inlet or conical entrance plate which has a bevelled leading edge and a very short cylindrical section. The orifices are normally concentric with the pipe (the eccentric orifice is a specific exception) and circular (except in the specific case of the segmental or chord orifice, in which the plate obstructs just a segment of the pipe). Standards and handbooks stipulate that the upstream surface of the plate is particularly flat and smooth. Sometimes a small drain or vent hole is drilled through the plate where it meets the pipe, to allow condensate or gas bubbles to pass along the pipe.

Pipe

Standards and handbooks stipulate a well-developed flow profile; velocities will be lower at the pipe wall than in the centre but not eccentric or jetting. Similarly the flow downstream of the plate must be unobstructed, otherwise the downstream pressure will be affected. To achieve this, the pipe must be acceptably circular, smooth and straight for stipulated distances. Sometimes when it is impossible to provide enough straight pipe, flow conditioners such as tube bundles or plates with multiple holes are inserted into the pipe to straighten and develop the flow profile, but even these require a further length of straight pipe before the orifice itself. Some standards and handbooks also provide for flows from or into large spaces rather than pipes, stipulating that the region before or after the plate is free of obstruction and abnormalities in the flow.

Theory

Incompressible Flow

By assuming steady-state, incompressible (constant fluid density), inviscid, laminar flow in a horizontal pipe (no change in elevation) with negligible frictional losses, Bernoulli's equation reduces to an equation relating the conservation of energy between two points on the same streamline:

$$p_1 + \frac{1}{2}\cdot\rho\cdot V_1'^2 = p_2 + \frac{1}{2}\cdot\rho\cdot V_2'^2$$

or:

$$p_1 - p_2 = \frac{1}{2}\cdot\rho\cdot V_2'^2 - \frac{1}{2}\cdot\rho\cdot V_1'^2$$

By continuity equation:

$$q_v' = A_1\cdot V_1' = A_2\cdot V_2' \quad \text{or} \quad V_1' = q_v' / A_1 \text{ and } V_2' = q_v' / A_2 :$$

$$p_1 - p_2 = \frac{1}{2}\cdot\rho\cdot\left(\frac{q'v}{A_2}\right)^2 - \frac{1}{2}\cdot\rho\cdot\left(\frac{q_v'}{A_1}\right)^2$$

Solving for q_v':

$$q_v' = A_2\sqrt{\frac{2\,(p_1 - p_2)/\rho}{1-(A_2/A_1)^2}}$$

and

$$q_v' = A_2 \sqrt{\frac{1}{1-(d/D)^4}} \sqrt{2(p_1 - p_2)/\rho}$$

The above expression for q_v' gives the theoretical volume flow rate. Introducing the beta factor $\beta = d/D$ as well as the discharge coefficient C_d:

$$q_v = C_d A_2 \sqrt{\frac{1}{1-\beta^4}} \sqrt{2(p_1 - p_2)/\rho}$$

And finally introducing the meter coefficient C which is defined as $C = \frac{C_d}{\sqrt{1-\beta^4}}$ to obtain the final equation for the volumetric flow of the fluid through the orifice which accounts for irreversible losses:

$$q_v = C A_2 \sqrt{2(p_1 - p_2)/\rho}$$

Multiplying by the density of the fluid to obtain the equation for the mass flow rate at any section in the pipe:

$$q_m = \rho q_v = C A_2 \sqrt{2\rho(p_1 - p_2)}$$

where:

- q_v = Volumetric flow rate (at any cross-section), m^3/s.
- q_v' = Theoretical volumetric flow rate (at any cross-section), m^3/s.
- q_m = Mass flow rate (at any cross-section), kg/s.
- q_m' = Theoretical mass flow rate (at any cross-section), kg/s.
- C_d = Coefficient of discharge, dimensionless.
- C = Orifice flow coefficient, dimensionless.
- A_1 = Cross-sectional area of the pipe, m^2.
- A_2 = Cross-sectional area of the orifice hole, m^2.
- D = Diameter of the pipe, m.
- d = Diameter of the orifice hole, m.
- β = Ratio of orifice hole diameter to pipe diameter, dimensionless.
- V_1' = Theoretical upstream fluid velocity, m/s.

- V_2' = Theoretical fluid velocity through the orifice hole, m/s.
- p_1 = Fluid upstream pressure, Pa with dimensions of kg/(m·s²).
- p_2 = Fluid downstream pressure, Pa with dimensions of kg/(m·s²).
- ρ = Fluid density, kg/m³.

Deriving the above equations used the cross-section of the orifice opening and is not as realistic as using the minimum cross-section at the vena contracta. In addition, frictional losses may not be negligible and viscosity and turbulence effects may be present. For that reason, the coefficient of discharge C_d is introduced. Methods exist for determining the coefficient of discharge as a function of the Reynolds number.

The parameter $\frac{1}{\sqrt{1-\beta^4}}$ is often referred to as the velocity of approach factor and multiplying the coefficient of discharge by that parameter (as was done above) produces the flow coefficient C. Methods also exist for determining the flow coefficient as a function of the beta function β and the location of the downstream pressure sensing tap. For rough approximations, the flow coefficient may be assumed to be between 0.60 and 0.75. For a first approximation, a flow coefficient of 0.62 can be used as this approximates to fully developed flow.

An orifice only works well when supplied with a fully developed flow profile. This is achieved by a long upstream length (20 to 40 pipe diameters, depending on Reynolds number) or the use of a flow conditioner. Orifice plates are small and inexpensive but do not recover the pressure drop as well as a venturi, nozzle, or venturi-nozzle does. Venturis also require much less straight pipe upstream. A venturi meter is more efficient, but usually more expensive and less accurate (unless calibrated in a laboratory) than an orifice plate.

Compressible Flow

In general, equation $q_m = \rho\, q_v = C\, A_2\, \sqrt{2\,\rho\,(p_1 - p_2)}$ is applicable only for incompressible flows. It can be modified by introducing the expansibility factor, (also called the expansion factor) ϵ to account for the compressibility of gasses.

$$q_m = \rho_1\, q_{v,1} = C\,\epsilon\, A_2\, \sqrt{2\,\rho_1\,(p_1 - p_2)}$$

ϵ is 1.0 for incompressible fluids and it can be calculated for compressible gases using empirically determined formulae as shown below in computation.

For smaller values of β (such as restriction plates with β less than 0.25 and discharge from tanks), if the fluid is compressible, the rate of flow depends on whether the flow has become choked. If it is, then the flow may be calculated as shown at choked flow (although the flow of real gases through thin-plate orifices never becomes fully choked).

By using a mechanical energy balance, compressible fluid flow in un-choked conditions may be calculated as:

$$q_m = C\,A_2 \sqrt{2\,\rho_1\,p_1 \left(\frac{}{\gamma-}\right)\left[(p_2/p_1)^{2/\gamma} - (p_2/p_1)^{(\gamma+1)/\gamma}\right]}$$

or

$$q_v = C\,A_2 \sqrt{2\,\frac{p_1}{\rho_1}\left(\frac{\gamma}{\gamma-1}\right)\left[(p_2/p_1)^{2/\gamma} - (p_2/p_1)^{(\gamma+1)/\gamma}\right]}$$

Under choked flow conditions, the fluid flow rate becomes:

$$q_m = C\,A_2 \sqrt{\gamma\,\rho_1\,p_1 \left(\frac{2}{\gamma+1}\right)^{\frac{\gamma+1}{\gamma-1}}}$$

or

$$q_v = C\,A_2 \sqrt{\gamma\,\frac{p_1}{\rho_1}\left(\frac{2}{\gamma+1}\right)^{\frac{\gamma+1}{\gamma-1}}}$$

where:

- γ = heat capacity ratio (c_p/c_v), dimensionless ($\gamma \approx 1.4$ for air)
- q_m, q_v = mass and volumetric flow rate, respectively, kg/s and m^3/s
- ρ_1 = real gas density under upstream conditions, kg/m^3.

Electromagnetic Flow Meters

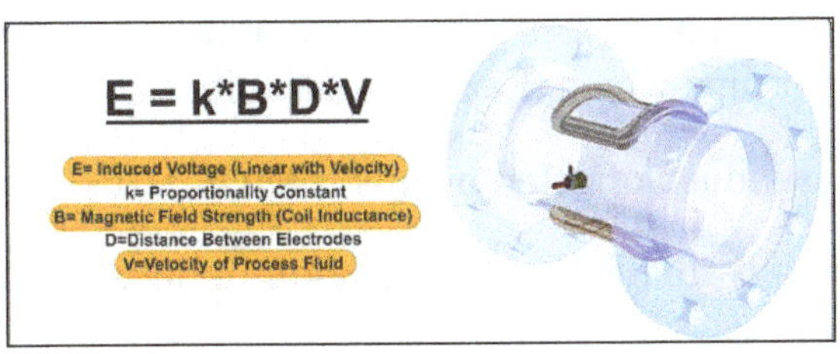

Electromagnetic flow meters are based on Faraday's law induction. These meters are also called as Magflow or Electromagnetic Flow Meters. A magnetic field is applied to the metering tube, which results in a potential difference proportional to the flow velocity perpendicular to the flux lines. The physical principle at work is electromagnetic induction and mathematically defined as E=k*B*D*V, where, E=Induced Voltage (Linear with velocity), k=Proportionality Constant, B=Magnetic Field Strength (Coil Inductance), D=Distance between electrodes, V=Velocity of process fluids.

The induced voltage (E) is directly proportional to the velocity (V) of the fluid moving through the magnetic field (B). The induced voltage is carried to the transmitter through the electrode circuit. The transmitter then converts this voltage into a quantifiable flow velocity. The volumetric flow rate of the fluid is calculated using this known velocity along with the area of the pipe.

When a flowmeter is installed and activated, its operations begin with a pair of charged magnetic coils. As energy passes through the coils, they produce a magnetic field that remains perpendicular to both the conductive fluid being measured and the axis of the electrodes taking measurements. The fluid moves along the longitudinal axis of the flowmeter, making any generated induced voltage perpendicular to the field and the fluid velocity. An increase in the flow rate of the conductive fluid will create a proportionate increase the voltage level.

The meter features flanged construction and is available with choice of liner and electrode material. All meters consist of a sensor and a converter that may be mounted integral to the sensor or remotely either with a field mount kit.

Industrial Applications

The Electromagnetic flow meter does not leave any obstructions in the pipe, so they are able to accommodate various applications from clean and conductive liquids. Electromagnetic flow meters are very well recommended for food grade applications like juices, milk, municipal or industrial water etc. These meters are also suitable for liquids were there are a lot of suspended particles like sewage, waste water, chemicals, and slurries etc.

Advantages of Electromagnetic Flow Meters

- The obstruction to the flow is almost nil and therefore this type of meters can be used for measuring heavy suspensions, including mud, sewage and wood pulp.
- There is no pressure head loss in this type of flow meter other than that of the length of straight pipe which the meter occupies.
- They are not very much affected by upstream flow disturbances.
- They are practically unaffected by variation in density, viscosity, pressure and temperature.
- Electric power requirements can be low (15 or 20 W), particularly with pulsed DC types.
- These meters can be used as bidirectional meters.
- The meters are suitable for most acids, bases, water and aqueous solutions because the lining materials selected are not only good electrical insulators but also are corrosion resistant.

- The meters are widely used for slurry services not only because they are obstruction less but also because some of the liners such as polyurethane, neoprene and rubber have good abrasion or erosion resistance.
- They are capable of handling extremely low flows.

Disadvantages of Electromagnetic Flow Meters

- These meters can be used only for fluids which have reasonable electrical conductivity.
- Accuracy is only in the range of ± 1% over a flow rate range of 5%.
- The size and cost of the field coils and circuitry do not increase in proportion to their size of pipe bore. Consequently small size meters are bulky and expensive.
- The substance being measured must be conductive. Therefore, it can't be employed for metering the flow rate of gases and steam, petroleum products and similar liquids having very low conductivity.
- To render the meter insensitive to variations in the resistance of liquid, the effective resistance of the liquid between the electrodes should not exceed 1% of the impedance of the external circuit.
- It is a very expensive device.
- As the meter always measures the volume rate, the volume of any suspended matter in the liquid will be included.
- To avoid any trouble which would be caused by entrained air, when the flow tube is installed in a horizontal pipe-line, the electrodes should be on the horizontal diameter.
- As a zero check on the installation can be performed only by stopping the flow, isolating valves are required and a bypass may also be necessary through which the flow may be directed during a zero check.
- The pipe must run full, in case regulating valves are installed upstream of the meter.

Ultrasonic Flow Meter

An ultrasonic flow meter is a type of flow meter that measures the velocity of a fluid with ultrasound to calculate volume flow. Using ultrasonic transducers, the flow meter can measure the average velocity along the path of an emitted beam of ultrasound, by averaging the difference in measured transit time between the pulses of ultrasound propagating into and against the direction of the flow or by measuring the frequency shift from the Doppler effect. Ultrasonic flow meters are affected by the acoustic properties of the fluid and can be impacted by temperature, density, viscosity and suspended particulates depending on the exact flow meter. They vary greatly in purchase price but are often inexpensive to use and maintain because they do not use moving parts, unlike mechanical flow meters.

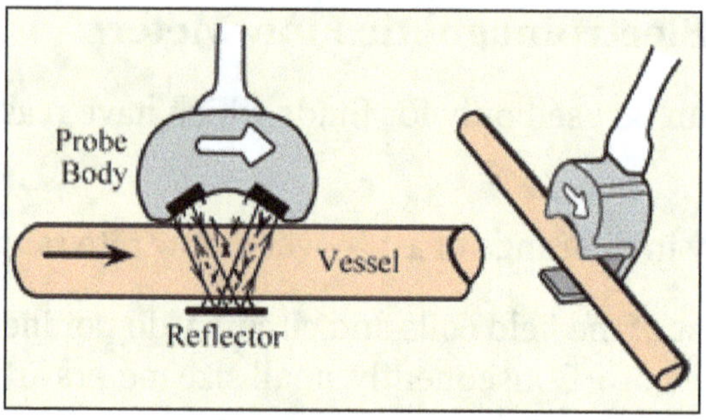

Schematic view of a flow sensor.

Means of Operation

There are three different types of ultrasonic flow meters. Transmission (or contrapropagating transit-time) flow meters can be distinguished into in-line (intrusive, wetted) and clamp-on (non-intrusive) varieties. Ultrasonic flow meters that use the Doppler shift are called reflection or Doppler flow meters. The third type is the open-channel flow meter.

Principle

Time Transit Flow Meter

Ultrasonic flow meters measure the difference between the transit time of ultrasonic pulses propagating with and against the flow direction. This time difference is a measure for the average velocity of the fluid along the path of the ultrasonic beam. By using the absolute transit times t_{up} and t_{down}, both the averaged fluid velocity v and the speed of sound c can be calculated. Using these two transit times, the distance between receiving and transmitting transducers L and the inclination angle α, if we assume that sound has to go against the flow when going up and along the flow when returning down, then one can write the following equations from the definition of velocity:

$$c - v\cos\alpha = \frac{L}{t_{up}} \text{ and } c + v\cos\alpha = \frac{L}{t_{down}}$$

By adding and subtracting the above equations we get,

$$v = \frac{L}{2\cos(\alpha)}\frac{t_{up} - t_{down}}{t_{up}\, t_{down}} \text{ and } c = \frac{L}{2}\frac{t_{up} + t_{down}}{t_{up}\, t_{down}}$$

where v is the average velocity of the fluid along the sound path and c is the speed of sound.

Doppler Shift Flow Meters

Another method in ultrasonic flow metering is the use of the Doppler shift that results from the reflection of an ultrasonic beam off sonically reflective materials, such as solid particles or entrained air bubbles in a flowing fluid, or the turbulence of the fluid itself, if the liquid is clean.

Doppler flowmeters are used for slurries, liquids with bubbles, gases with sound-reflecting particles.

This type of flow meter can also be used to measure the rate of blood flow, by passing an ultrasonic beam through the tissues, bouncing it off a reflective plate, then reversing the direction of the beam and repeating the measurement, the volume of blood flow can be estimated. The frequency of the transmitted beam is affected by the movement of blood in the vessel and by comparing the frequency of the upstream beam versus downstream the flow of blood through the vessel can be measured. The difference between the two frequencies is a measure of true volume flow. A wide-beam sensor can also be used to measure flow independent of the cross-sectional area of the blood vessel.

Open Channel Flow Meters

In this case, the ultrasonic element is actually measuring the height of the water in the open channel; based on the geometry of the channel, the flow can be determined from the height. The ultrasonic sensor usually also has a temperature sensor with it because the speed of sound in air is affected by the temperature.

References

- Different-types-of-tachogenerators-that-we-offer: tachogenerators.co.uk, Retrieved 10 July, 2020
- About-inductive-transducers-62717: brighthubengineering.com, Retrieved 05 January, 2020
- Capacitive-transducer: circuitglobe.com, Retrieved 25 June, 2020
- Lvdt-construction-and-working-principle: electricalbaba.com, Retrieved 28 August, 2020
- Rvdt-advantages-and-common-applications: sdgyoungleaders.org, Retrieved 04 March, 2020
- Electromagnetic-Flow-Meters-Working-Principle: eeplindia.com, Retrieved 08 April, 2020
- Lynnworth, L.C.: Ultrasonic Measurements for Process Control. Academic Press, Inc. San Diego. ISBN 0-12-460585-0

5

Robotics

The interdisciplinary branch of engineering and science that is concerned with the design, operation, construction, and use of robots is referred to as robotics. It has a wide range of applications in the fields of medicine, agriculture, engineering, navigation and households. This chapter closely examines the field of robotics to provide an extensive understanding of the subject.

Robotics is a branch of engineering and science that includes electronics engineering, mechanical engineering and computer science and so on. This branch deals with the design, construction, use to control robots, sensory feedback and information processing. These are some technologies which will replace humans and human activities in coming years. These robots are designed to be used for any purpose but these are using in sensitive environments like bomb detection, deactivation of various bombs etc. Robots can take any form but many of them have given the human appearance. The robots which have taken the form of human appearance may likely to have the walk like humans, speech, cognition and most importantly all the things a human can do. Most of the robots of today are inspired by nature and are known as bio-inspired robots. Characteristics.

There are some characteristics of robots given below:

- Appearance: Robots have a physical body. They are held by the structure of their body and are moved by their mechanical parts. Without appearance, robots will be just a software program.
- Brain: Another name of brain in robots is on-board control unit. Using this robot receive information and sends commands as output. With this control unit robot knows what to do else it'll be just a remote-controlled machine.
- Sensors: The use of these sensors in robots is to gather info from the outside world and send it to Brain. Basically, these sensors have circuits in them that produces the voltage in them.
- Actuators: The robots move and the parts with the help of these robots move

is called Actuators. Some examples of actuators are motors, pumps, and compressor etc. The brain tells these actuators when and how to respond or move.

- Program: Robots only works or responds to the instructions which are provided to them in the form of a program. These programs only tell the brain when to perform which operation like when to move, produce sounds etc. These programs only tell the robot how to use sensors data to make decisions.
- Behaviour: Robots behavior is decided by the program which has been built for it. Once the robot starts making the movement, one can easily tell which kind of program is being installed inside the robot.

Robots

A robot is a machine designed to execute one or more tasks automatically with speed and precision. There are as many different types of robots as there are tasks for them to perform. Robots that resemble humans are known as androids; however, many robots aren't built on the human model. Industrial robots, for example, are often designed to perform repetitive tasks that aren't facilitated by a human-like construction. A robot can be remotely controlled by a human operator, sometimes from a great distance. A telechir is a complex robot that is remotely controlled by a human operator for a telepresence system, which gives that individual the sense of being on location in a remote, dangerous or alien environment and the ability to interact with it. Telepresence robots, which simulate the experience and some of the capabilities of being physically present, can enable remote business consultations, healthcare, home monitoring and childcare, among many other possibilities.

An autonomous robot acts as a stand-alone system, complete with its own computer (called the controller). The most advanced example is the smart robot, which has a built-in artificial intelligence (AI) system that can learn from its environment and its experience and build on its capabilities based on that knowledge.

Swarm robots, sometimes referred to as insect robots, work in fleets ranging in number from a few to thousands, with all fleet members under the supervision of a single controller. The term arises from the similarity of the system to a colony of insects, where the individuals and behaviors are simple but the fleet as a whole can be sophisticated.

Robots are sometimes grouped according to the time frame in which they were first widely used. First-generation robots date from the 1970s and consist of stationary, nonprogrammable, electromechanical devices without sensors. Second-generation robots were developed in the 1980s and can contain sensors and programmable controllers. Third-generation robots were developed between approximately 1990 and the present. These machines can be stationary or mobile, autonomous or insect type, with sophisticated programming, speech recognition and/or synthesis, and other advanced features. Fourth-generation robots are in the research-and-development phase, and include features such as artificial intelligence, self-replication, self-assembly, and nanoscale size (physical dimensions on the order of nanometers, or units of 10^{-9} meter).

Some advanced robots are called androids because of their superficial resemblance to human beings. Androids are mobile, usually moving around on wheels or a track drive (robots legs are unstable and difficult to engineer). The android is not necessarily the end point of robot evolution. Some of the most esoteric and powerful robots do not look or behave anything like humans. The ultimate in robotic intelligence and sophistication might take on forms yet to be imagined.

Components of Robotics

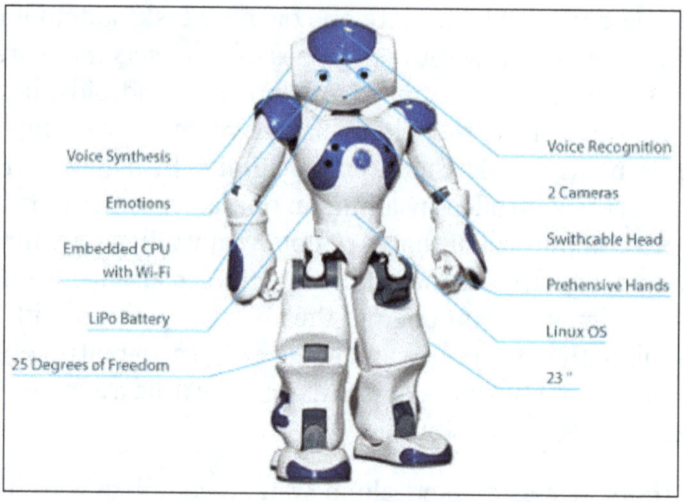

Consider the key components of robotics are:

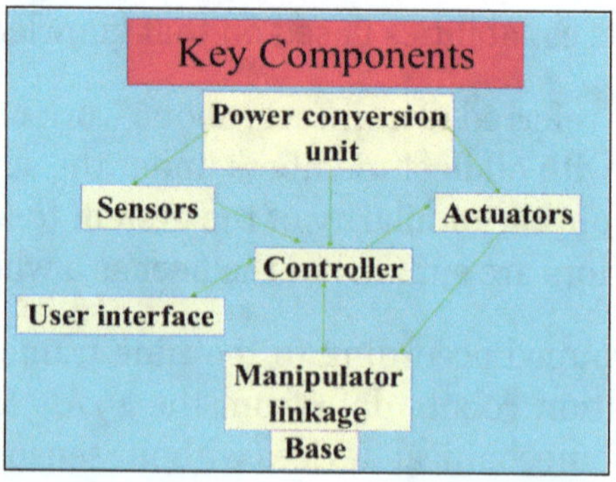

- Power Supply: The working power to the robot is provided by batteries, hydraulic, solar power, or pneumatic power sources.

- Actuators: Actuators are the energy conversion device used inside a robot. The major function of actuators is to convert energy into movement.
- Electric motors (DC/AC): Motors are electromechanical component used for converting electrical energy into its equivalent mechanical energy. In robots motors are used for providing rotational movement.
- Sensors: Sensors provide real time information on the task environment. Robots are equipped with tactile sensor it imitates the mechanical properties of touch receptors of human fingerprints and a vision sensor is used for computing the depth in the environment.
- Controller: Controller is a part of robot that coordinates all motion of the mechanical system. It also receives an input from immediate environment through various sensors. The heart of robot's controller is a microprocessor linked with the input/output and monitoring device. The command issued by the controller activates the motion control mechanism, consisting of various controller, actuators and amplifier.

Types of Robots

Humanoid

A humanoid robot is a robot with its body shape built to resemble the human body. The design may be for functional purposes, such as interacting with human tools and environments, for experimental purposes, such as the study of bipedal locomotion, or for other purposes. In general, humanoid robots have a torso, a head, two arms, and two legs, though some forms of humanoid robots may model only part of the body, for example, from the waist up. Some humanoid robots also have heads designed to replicate human facial features such as eyes and mouths. Androids are humanoid robots built to aesthetically resemble humans.

Purpose

Humanoid robots are now used as research tools in several scientific areas. Researchers study the human body structure and behavior (biomechanics) to build humanoid robots. On the other side, the attempt to simulate the human body leads to a better understanding of it. Human cognition is a field of study which is focused on how humans learn from sensory information in order to acquire perceptual and motor skills. This knowledge is used to develop computational models of human behavior and it has been improving over time.

It has been suggested that very advanced robotics will facilitate the enhancement of ordinary humans.

Although the initial aim of humanoid research was to build better orthosis and prosthesis for human beings, knowledge has been transferred between both disciplines. A few examples are powered leg prosthesis for neuromuscularly impaired, ankle-foot orthosis, biological realistic leg prosthesis and forearm prosthesis.

Valkyrie, from NASA.

Besides the research, humanoid robots are being developed to perform human tasks like personal assistance, through which they should be able to assist the sick and elderly, and dirty or dangerous jobs. Humanoids are also suitable for some procedurally-based vocations, such as reception-desk administrators and automotive manufacturing line workers. In essence, since they can use tools and operate equipment and vehicles designed for the human form, humanoids could theoretically perform any task a human being can, so long as they have the proper software. However, the complexity of doing so is immense.

They are also becoming increasingly popular as entertainers. For example, Ursula, a female robot, sings, plays music, dances and speaks to her audiences at Universal Studios. Several Disney theme park shows utilize animatronic robots that look, move and speak much like human beings. Although these robots look realistic, they have no cognition or physical autonomy. Various humanoid robots and their possible applications in daily life are featured in an independent documentary film called *Plug & Pray*, which was released in 2010.

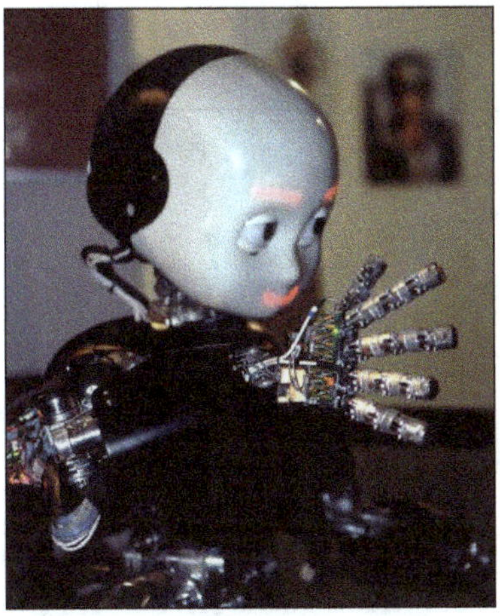

iCub robot.

Humanoid robots, especially those with artificial intelligence algorithms, could be useful for future dangerous and/or distant space exploration missions, without having the need to turn back around again and return to Earth once the mission is completed.

Sensors

A sensor is a device that measures some attribute of the world. Being one of the three primitives of robotics (besides planning and control), sensing plays an important role in robotic paradigms.

Sensors can be classified according to the physical process with which they work or according to the type of measurement information that they give as output. In this case, the second approach was used.

Proprioceptive

Proprioceptive sensors sense the position, the orientation and the speed of the humanoid's body and joints.

In human beings the otoliths and semi-circular canals (in the inner ear) are used to maintain balance and orientation. In addition humans use their own proprioceptive sensors (e.g. touch, muscle extension, limb position) to help with their orientation. Humanoid robots use accelerometers to measure the acceleration, from which velocity can be calculated by integration; tilt sensors to measure inclination; force sensors placed in robot's hands and feet to measure contact force with environment; position sensors, that indicate the actual position of the robot (from which the velocity can be calculated by derivation) or even speed sensors.

Exteroceptive

An artificial hand holding a lightbulb.

Arrays of tactels can be used to provide data on what has been touched. The Shadow Hand uses an array of 34 tactels arranged beneath its polyurethane skin on each finger tip. Tactile sensors also provide information about forces and torques transferred between the robot and other objects.

Vision refers to processing data from any modality which uses the electromagnetic spectrum to produce an image. In humanoid robots it is used to recognize objects and determine their properties. Vision sensors work most similarly to the eyes of human beings. Most humanoid robots use CCD cameras as vision sensors.

Sound sensors allow humanoid robots to hear speech and environmental sounds, and perform as the ears of the human being. Microphones are usually used for this task.

Actuators

Actuators are the motors responsible for motion in the robot.

Humanoid robots are constructed in such a way that they mimic the human body, so they use actuators that perform like muscles and joints, though with a different structure. To achieve the same effect as human motion, humanoid robots use mainly rotary actuators. They can be either electric, pneumatic, hydraulic, piezoelectric or ultrasonic.

Hydraulic and electric actuators have a very rigid behavior and can only be made to act in a compliant manner through the use of relatively complex feedback control strategies. While electric coreless motor actuators are better suited for high speed and low load applications, hydraulic ones operate well at low speed and high load applications.

Piezoelectric actuators generate a small movement with a high force capability when voltage is applied. They can be used for ultra-precise positioning and for generating and handling high forces or pressures in static or dynamic situations.

Ultrasonic actuators are designed to produce movements in a micrometer order at ultrasonic frequencies (over 20 kHz). They are useful for controlling vibration, positioning applications and quick switching.

Pneumatic actuators operate on the basis of gas compressibility. As they are inflated, they expand along the axis, and as they deflate, they contract. If one end is fixed, the other will move in a linear trajectory. These actuators are intended for low speed and low/medium load applications. Between pneumatic actuators there are: cylinders, bellows, pneumatic engines, pneumatic stepper motors and pneumatic artificial muscles.

Planning and Control

In planning and control, the essential difference between humanoids and other kinds of robots (like industrial ones) is that the movement of the robot must be human-like, using legged locomotion, especially biped gait. The ideal planning for humanoid movements during normal walking should result in minimum energy consumption, as it does in the human body. For this reason, studies on dynamics and control of these kinds of structures has become increasingly important.

The question of walking biped robots stabilization on the surface is of great importance. Maintenance of the robot's gravity center over the center of bearing area for providing a stable position can be chosen as a goal of control.

To maintain dynamic balance during the walk, a robot needs information about contact force and its current and desired motion. The solution to this problem relies on a major concept, the Zero Moment Point (ZMP).

Another characteristic of humanoid robots is that they move, gather information (using sensors) on the "real world" and interact with it. They don't stay still like factory manipulators and other robots that work in highly structured environments. To allow humanoids to move in complex environments, planning and control must focus on self-collision detection, path planning and obstacle avoidance.

Humanoid robots do not yet have some features of the human body. They include structures with variable flexibility, which provide safety (to the robot itself and to the people), and redundancy of movements, i.e. more degrees of freedom and therefore wide task availability. Although these characteristics are desirable to humanoid robots, they will bring more complexity and new problems to planning and control. The field of whole-body control deals with these issues and addresses the proper coordination of numerous degrees of freedom, e.g. to realize several control tasks simultaneously while following a given order of priority.

Domestic Robots

A domestic robot is a type of service robot, an autonomous robot that is primarily used for household chores, but may also be used for education, entertainment or therapy. Thus far, there are only a few limited models, though speculators, such as Bill Gates, have suggested that they could become more common in the future. While most domestic robots are simplistic, some are connected to WiFi home networks or smart environments and are autonomous to a high degree. There were an estimated 3,540,000 service robots in use in 2006, compared with an estimated 950,000 industrial robots.

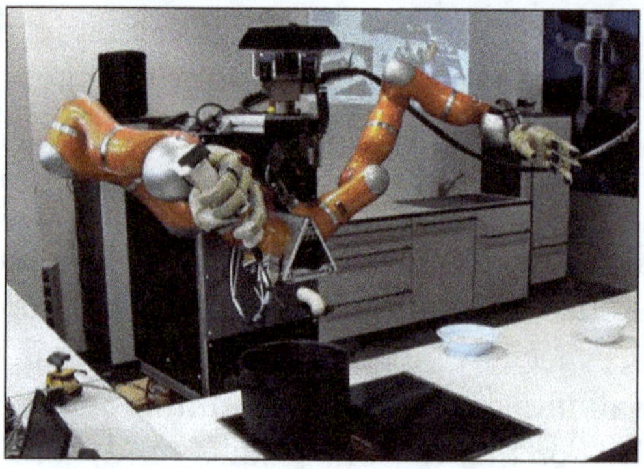

Indoor Robots

This type of domestic robot does chores around and inside homes. Different kinds include:

Robotic vacuum cleaners and floor-washing robots that clean floors with sweeping and wet mopping functions. Some use Swiffer or other disposable cleaning cloths to dry-sweep, or reusable microfiber cloths to wet-mop.

Within the ironing robots, Dressman is a mannequin to dry and iron shirts using hot air. Other ones also includes mannequin for down parts (pants, trousers and skirts). More advanced ones fold and organizes the clothes, as Laundroid (using image analysis and artificial intelligence), Effie (irons 12 items of clothing at once) and FoldiMate.

Cat litter robots are automatic self-cleaning litter boxes that filter clumps out into a built-in waste receptacle that can be lined with an ordinary plastic bag.

Robotic kitchens include Rotimatic (which makes rotis, tortillas, puris out of flour in just a few minutes), Moley Robotics MK1 and Prometheus delta robot.

Security robots such as Knightscope have a night-vision-capable wide-angle camera that detects movements and intruders. It can patrol places and shoot video of suspicious activities, too, and send alerts via email or text message; the stored history of past alerts and videos are accessible via the Web. The robot can also be configured to go into action at any time of the day.

Atlas is a robot built to perform in house task such as sweeping, opening doors, climbing stairs, etc.

Outdoor Robots

A robotic lawn mower is a lawn mower that is able to mow a lawn by itself after being programmed. Once programmed, this invention repeats the operation by itself according to its programming. Robotic lawn mowers come with a power unit which may be an electric motor or internal combustion engine. This provides power to the robot and allows it to move itself and its cutting blades. There is also a control unit which helps the mower move. This unit also contains a memory unit which records and memorizes its operation programming. Its memorized route includes the length of travel in a given direction and turns angles. This allows the same lawn to be mowed repeatedly without having to reprogram. The steering unit acquires an operation signal and propels the lead wheel, which leads the mower, go guide along the programmed route.

Some models can mow complicated and uneven lawns that are up to three-quarters of an acre in size. Others can mow a lawn as large as 40,000 square feet (3,700 m^2), can handle a hill inclined up to 27 degrees.

There are also automated pool cleaners that clean and maintain swimming pools autonomously by scrubbing in-ground pools from the floor to the waterline in 3 hours, cleaning and circulating more than 70 US gallons (260 l) of water per minute, and removing debris as small as 2 μm in size.

Gutter-cleaning robots such as Looj use brushes and rubber blades to remove debris from rain gutters; users operate the device using a remote.

Window cleaning robots are most commonly used to clean outdoor windows, more specifically house windows. However, it may be used on other types of windows, such as ones on tall buildings and structures. This robot contains a movement system which allows the robot to navigate itself across the window surface in a defined direction. It also has a powered agitator located by the cleaning pad. When activated, the agitator gets rid of debris and dirt from the window surface. The cleaning pad directly interacts with the window surface and is directly responsible for removing the dirt by filling itself with specialized window cleaning fluid.

A window-washing robot commonly uses two magnetic modules to navigate windows as it sprays the cleaning solution onto microfiber pads to wash them. It covers about 1,601 square feet (148.7 m^2) per charge.

Toys

Robotic toys, such as the well known Furby, have been popular since 1998. There are also small humanoid remote-controlled robots as well as electronic pets, such as robotic dogs. They have also have been used by many universities in competitions such as the RoboCup.

Social Robots

Social robots take on the function of social communication. Domestic humanoid robots are used by elderly and immobilized residents to keep them company.

Home-telepresence robots can move around in a remote location and let one communicate with people there via its camera, speaker, and microphone.

Network robots link ubiquitous networks with robots, contributing to the creation of new lifestyles and solutions to address a variety of social problems including the aging of population and nursing care.

Robots built for therapy have been in production for quite some time now. Some of these uses can be for autism or physical therapy.

Robots in Entertainment

An entertainment robot is, as the name indicates, a robot that is not made for utilitarian use, as in production or domestic services, but for the sole subjective pleasure of the human. It serves, usually the owner or his housemates, guests or clients. Robotics technologies are applied in many areas of culture and entertainment.

Expensive robotics are applied to the creation of narrative environments in commercial venues where servo motors, pneumatics and hydraulic actuators are used to create movement with often preprogrammed responsive behaviors such as in Disneyland's haunted house ride.

Entertainment robots can also be seen in the context of media arts where artist have been employing advanced technologies to create environments and artistic expression also utilizing the actuators and sensor to allow their robots to react and change in relation to viewers.

Toy Robot

Relatively cheap, mass-produced entertainment robots are used as mechanical, sometimes interactive, toys which perform various tasks and tricks on command. The first commercial hit was, not surprisingly, modelled on the most popular pet: the canine.

Robotic Dog

Robot dogs as a fad have been produced with relatively little variation. These are some commercial models:

- Teksta a toy robot dog popular in the 1990s which was intended to be able to perform card tricks and respond to commands.
- Aibo (The Sony robot dog).

- Poo-Chi.
- Bo-wow.
- I-Cybie.
- iDog (Sega's robot iPod music speaker).
- Gupi, a robotic guinea pig.
- Space Dog, the remote control dog.

Robot dogs also appear fairly frequently in fiction compared to other forms of personal entertainment robots.

- K-9: The Doctor's portable computer and robot, from the British BBC Television series *Doctor Who*.
- Preston: Wendolene's robot dog from the 1995 animated Wallace and Gromit film *A Close Shave*.
- Goddard pet of Jimmy Neutron.

Humanoid Entertainment Robots

Despite those humanoid robots for utilitarian uses, there are some humanoid robots which aims at entertainment uses, such as Sony's QRIO and Wow Wee's RoboSapien. They are usually capable of some advanced features like Voice Recognition or Walking.

Substitute Pets

While primitive robot toy models only execute standardized pre-programmed routines, sometimes little more than a wind-up toy could do, advancing technology allows for interaction with the user and/or other environmental stimuli (e.g. sensor-detected obstacles), thus somewhat resembling a live playmate.

Nevertheless, in the mind of some users the things can hold the loved place of a pet, as demonstrated by the fact that some even sleep with a metallic one instead of a plush cuddly toy.

In fact manufacturers even found it pays to produce a toy that is essentially designed to be nurtured, rather like an egg in some 'parenting experience simulations', as proven by the success of the Japanese Tamagotchi.

Entertainment robots can take the form of interactive communications marketing tools at trade shows wherein promotional robots move about a trade show floor providing tongue-in-cheek interaction with trade show attendees in order to bring said attendees to a particular companies trade show booth. These promotional robots are hired by companies to entertain the inform trade show attendees about products ad services available at trade shows.

Commercial Show Robots

As usual in the entertainment industry, capital and creativity are invested to try and top anything the private person can afford. In fact, from their owner's point of view this is a professional use, but the product is designed with as end use in mind its appreciation by the public.

Thus expensive robots are made for use as:

- Marketing tool: Logically showed off by the manufacturers, to promote their products and technology, occasionally used in other promotional productions.
- Prop: Inanimate performer or even artificial actor in show, tv and movie production; as technology advances, some advanced robots can, often helped with other special effects, to make them seem what cannot (yet), even be significantly more than a cast extra, such as the droids r2-d2 and c-3po in the *star wars* double trilogy which have proved rather popular from the start.

Non-commercial Art Robots

In 1956, Nicolas Schöffer created *Cysp 1 (Spatiodynamique Cybernétique)*, a robot and dancer working together to create an abstract sculpture and choreography with concrete music by Pierre Henry. These works could react to color, sound and light.

Emergent Systems is creating large-scale interactive art environments where robots are able to respond to humans and each other as they react and evolve in the robotic installations. Autopoiesis was one such artificial life work that allowed a series of robots constructed of grapevines to both act as individuals and a group. Augmented Fish Reality allowed Siamese fighting fish to control their robots to meet across the gap of their glass fish bowls.

Intel Museum hosts the A.I. driven interactive robot, ARTI, which is short for "artificial intelligence". This robot is considered to be a work of fine art and is capable of recognizing faces, understands speech and even teaches the museum guests about the history of the museum and its founders, Robert Noyes and Gordon Moore. ARTI's face is made out of an inanimate silicon wafer.

Robots in Military

Military robots are autonomous robots or remote-controlled mobile robots designed for military applications, from transport to search & rescue and attack.

Some such systems are currently in use, and many are under development.

Armed Predator drone.

Broadly defined, military robots date back to World War II and the Cold War in the form of the German Goliath tracked mines and the Soviet teletanks. The MQB-1 Predator drone was when "CIA officers began to see the first practical returns on their decade-old fantasy of using aerial robots to collect intelligence".

The use of robots in warfare, although traditionally a topic for science fiction, is being researched as a possible future means of fighting wars. Already several military robots have been developed by various armies. Some believe the future of modern warfare will be fought by automated weapons systems. The U.S. military is investing heavily in research and development towards testing and deploying increasingly automated systems. The most prominent system currently in use is the unmanned aerial vehicle (IAI Pioneer & RQ-1 Predator) which can be armed with air-to-ground missiles and remotely operated from a command center in reconnaissance roles. DARPA has hosted competitions in 2004 & 2005 to involve private companies and universities to develop unmanned ground vehicles to navigate through rough terrain in the Mojave Desert for a final prize of 2 million.

Artillery has seen promising research with an experimental weapons system named "Dragon Fire II" which automates loading and ballistics calculations required for accurate predicted fire, providing a 12-second response time to fire support requests. However, military weapons are prevented from being fully autonomous; they require human input at certain intervention points to ensure that targets are not within restricted fire areas as defined by Geneva Conventions for the laws of war.

There have been some developments towards developing autonomous fighter jets and bombers. The use of autonomous fighters and bombers to destroy enemy targets is especially promising because of the lack of training required for robotic pilots, autonomous planes are capable of performing maneuvers which could not otherwise be done with human pilots (due to high amount of G-force), plane designs do not require a life support system, and a loss of a plane does not mean a loss of a pilot. However, the largest drawback to robotics is their inability to accommodate for non-standard conditions. Advances in artificial intelligence in the near future may help to rectify this.

In Current Use

Foster-Miller TALON SWORDS units equipped with various weaponry.

The Platforma-M variant of the Multifunctional Utility/Combat support/ Patrol. Serially produced by the Russian Army.

- DRDO Daksh.
- D9T Panda, Israel.
- Elbit Hermes 450, Israel.
- Goalkeeper CIWS.
- Guardium.
- IAIO Fotros, Iran.
- PackBot.
- MQ-9 Predator B.
- MQ-1 Predator.
- TALON.
- Samsung SGR-A1.
- Shahed 129, Iran.
- Shomer Gvouloth ("Border Keeper").

In Development

The Armed Robotic Vehicle variant of the MULE.

- US Mechatronics has produced a working automated sentry gun and is currently developing it further for commercial and military use.
- MIDARS, a four-wheeled robot outfitted with several cameras, radar, and possibly a firearm, that automatically performs random or preprogrammed patrols around a military base or other government installation. It alerts a human overseer when it detects movement in unauthorized areas, or other programmed conditions. The operator can then instruct the robot to ignore the event, or take over remote control to deal with an intruder, or to get better camera views of an emergency. The robot would also regularly scan radio frequency identification tags (RFID) placed on stored inventory as it passed and report any missing items.
- Tactical Autonomous Combatant (TAC) units, described in Project Alpha study Unmanned Effects: Taking the Human out of the Loop.
- Autonomous Rotorcraft Sniper System is an experimental robotic weapons system being developed by the U.S. Army since 2005. It consists of a remotely operated sniper rifle attached to an unmanned autonomous helicopter. It is intended for use in urban combat or for several other missions requiring snipers. Flight tests are scheduled to begin in summer 2009.
- The "Mobile Autonomous Robot Software" research program was started in December 2003 by the Pentagon who purchased 15 Segways in an attempt to develop more advanced military robots. The program was part of a $26 million Pentagon program to develop software for autonomous systems.
- ACER.
- Atlas (robot).
- Battlefield Extraction-Assist Robot.
- Dassault nEUROn (French UCAV).

- Dragon Runner.
- MATILDA.
- MULE (US UGV).
- R-Gator.
- Ripsaw MS1.
- SUGV.
- Syrano.
- iRobot Warrior.
- PETMAN.
- Excalibur unmanned aerial vehicle.

Effects and Impact

Advantages

Increasing attention is also paid to how to make the robots more autonomous, with a view of eventually allowing them to operate on their own for extended periods of time, possibly behind enemy lines. For such functions, systems like the Energetically Autonomous Tactical Robot are being tried, which is intended to gain its own energy by foraging for plant matter. The majority of military robots are tele-operated and not equipped with weapons; they are used for reconnaissance, surveillance, sniper detection, neutralizing explosive devices, etc. Current robots that are equipped with weapons are tele-operated so they are not capable of taking lives autonomously. Advantages regarding the lack of emotion and passion in robotic combat is also taken into consideration as a beneficial factor in significantly reducing instances of unethical behavior in wartime. Autonomous machines are created not to be "truly 'ethical' robots", yet ones that comply with the laws of war (LOW) and rules of engagement (ROE). Hence the fatigue, stress, emotion, adrenaline, etc. that affect a human soldier's rash decisions are removed; there will be no effect on the battlefield caused by the decisions made by the individual.

Risks

Human rights groups and NGOs such as Human Rights Watch and the Campaign to Stop Killer Robots have started urging governments and the United Nations to issue policy to outlaw the development of so-called "lethal autonomous weapons systems" (LAWS). The United Kingdom opposed such campaigns, with the Foreign Office declaring that "international humanitarian law already provides sufficient regulation for this area".

In July 2015, over 1,000 experts in artificial intelligence signed a letter calling for a ban on autonomous weapons. The letter was presented in Buenos Aires at the 24th International Joint Conference on Artificial Intelligence (IJCAI-15) and was co-signed by Stephen Hawking, Elon Musk, Steve Wozniak, Noam Chomsky, Skype co-founder Jaan Tallinn and Google DeepMind co-founder Demis Hassabis, among others.

Psychology

American soldiers have been known to name the robots that serve alongside them. These names are often in honor of human friends, family, celebrities, pets, or are eponymic. The 'gender' assigned to the robot may be related to the marital status of its operator.

Some affixed fictitious medals to battle-hardened robots, and even held funerals for destroyed robots. An interview of 23 explosive ordnance detection members shows that while they feel it is better to lose a robot than a human, they also felt anger and a sense of loss if they were destroyed. A survey of 746 people in the military showed that 80% either 'liked' or 'loved' their military robots, with more affection being shown towards ground rather than aerial robots. Surviving dangerous combat situations together increased the level of bonding between soldier and robot, and current and future advances in artificial intelligence may further intensify the bond with the military robots.

Robots in Medicine

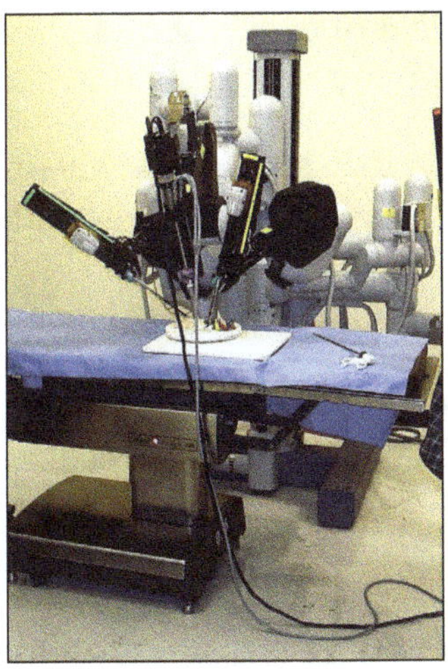

A medical robot is a robot used in the medical sciences. They include surgical robots. These are in most telemanipulators, which use the surgeon's activators on one side to control the "effector" on the other side.

Types

- Surgical robots: Either allow surgical operations to be carried out with big precision than an unaided human surgeon or allow remote surgery where a human surgeon is not physically present with the patient.

- Rehabilitation robots: Facilitate and support the lives of infirm, elderly people, or those with dysfunction of body parts affecting movement. These robots are also used for rehabilitation and related procedures, such as training and therapy.
- Biorobots: A group of robots designed to imitate the cognition of humans and animals.
- Telepresence robots: Allow off-site medical professionals to move, look around, communicate, and participate from remote locations.
- Pharmacy automation: Robotic systems to dispense oral solids in a retail pharmacy setting or preparing sterile IV admixtures in a hospital pharmacy setting.
- Companion robot: Has the capability to engage emotionally with users keeping them company and alerting if there is a problem with their health.
- Disinfection robot: Has the capability to disinfect a whole room in mere minutes, generally using pulsed ultraviolet light. They are being used to fight Ebola virus disease.

Robots for Disabled Persons

A disability robot is a robot designed to help people who have physical disabilities that impede with daily tasks. The field of expertise that creates such robots is called "disability robotics".

Disability robot has been proven to assist people who are recovering from strokes and people who have abstained injuries that effect their daily tasks.

Gadgets

Disability robotics is a broad category that includes wheelchairs, robotic arms, and other robotic devices that assist disabled persons of all ability levels.

Wheelchair

Persons with severe disabilities may be assisted with robotic wheelchairs when manual control is not possible. These devices can deter loss of residual skills and frustration. Traditionally wheelchairs either gave control to the person or robot depending on disability level.

Treadmill

Bodyweight-supported treadmill training (BWSTT) are used to enhance walking ability of people with neurological injury. These machines are therapist-assisted devices that are used in the clinical setting, but is limited by the personnel and labor requirements placed on physical therapists. The BWSTT device, and many others like it, assist physical therapists by providing task-specific practice of walking in people following neurological injury.

Applications of Robotics

The automotive manufacturing industry has long been one of the quickest and largest adopters of industrial robotic technology, and that continues to this day. Robots are used in nearly every part of automotive manufacturing in one way or another, and it remains as one of the most highly automated supply chains in the world.

While there are plenty of robotic applications to choose from within the industry, there are 6 that stand out as the most common and most valuable applications on the market.

The following robotic applications are the most common in the automotive industry:

- Collaborative Robots: These collaborative robots are built to work together with other robots, on enormous assembly lines. Robots must collaborate between handling and welding robots to make such assembly lines function properly.
- Robotic Painting: Professional painters are difficult to find and the job is a highly toxic one. This makes it perfect for robots, because the paint job needs to be highly consistent over a large area of paint, and reducing the amount of wasted material can add up to quite a bit of savings over time.
- Robotic Welding: Robotic welding has been the top robotic application in the automotive sector for a long time, as every car needs a high number of welds before it's complete. Given the high value of the finished product, productivity from automation is enormous.

- Robotic Assembly: In many automotive plants, robots are assembling smaller components like pumps and motors at high speeds. Often, robots are performing tasks like windshield installation and wheel mounting to increase throughput.
- Material Removal: High consistency and repeatability make robots perfect for material removal processes like trimming and cutting. This could be in the form of cutting fabrics, trimming plastic moldings and die castings or even polishing molds.
- Part Transfer and Machine Tending: Pouring molten metal, transferring metal stamps, and loading and unloading CNC machines are all best completed by a robot as they are dangerous. When completed consistently with little downtime they can also be a source of major productivity.

References

- Robotics-introduction: geeksforgeeks.org, Retrieved 17 May, 2020
- Robot, definition: searchenterpriseai.techtarget.com, Retrieved 19 July, 2020
- Components-of-robot: javatpoint.com, Retrieved 06 January, 2020
- Hellström, Thomas (June 2013). "On the moral responsibility of military robots". Ethics and Information Technology. 15 (2): 99–107. CiteSeerX 10.1.1.305.5964. doi:10.1007/s10676-012-9301-2
- Robots-automotive-manufacturing-top-6-applications: genesis-systems.com, Retrieved 14 June, 2020
- Beasley, Ryan A. (12 August 2012). "Medical Robots: Current Systems and Research Directions". Journal of Robotics. 2012: 1–14. doi:10.1155/2012/401613
- Sivan, Gallagher, Holt, Weightman, O'Connor, Levesley, Manoj, Justin, Ray, Andrew, Rory, Martin (6 February 2016). "Employing the International Classification of Functioning, Disability and Health framework to capture user feedback in the design and testing stage of development of home-based arm rehabilitation technology". Assistive Technology. 28 (3): 175–182. doi:10.1080/10400435.2016.1140689
- Gates, William 'Bill' III (January 2007). "A Robot in Every Home". Scientific American. 296 (1): 58–65. Bibcode:2007SciAm.296a..58G. doi:10.1038/scientificamerican0107-58

6
Control System

A type of system that directs, commands, manages and regulates the behavior of different systems through control loops is termed as control system. It includes open and closed loop systems, feedback systems, PID control, etc. This chapter has been carefully written to provide an easy understanding of control system.

Classification of Control System

A control system is a system, which provides the desired response by controlling the output. The following figure shows the simple block diagram of a control system.

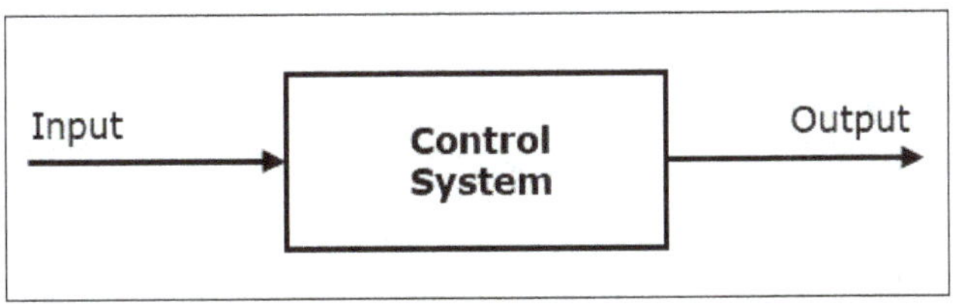

Here, the control system is represented by a single block. Since, the output is controlled by varying input, the control system got this name. We will vary this input with some mechanism. In the next section on open loop and closed loop control systems, we will study in detail about the blocks inside the control system and how to vary this input in order to get the desired response.

Examples: Traffic lights control system, washing machine.

Traffic lights control system is an example of control system. Here, a sequence of input signal is applied to this control system and the output is one of the three lights that will be on for some duration of time. During this time, the other two lights will be off. Based on the traffic study at a particular junction, the on and off times of the lights can be determined. Accordingly, the input signal controls the output. So, the traffic lights control system operates on time basis.

Based on some parameters, we can classify the control systems into the following ways:

- Continuous time and discrete-time control systems:
  - Control Systems can be classified as continuous time control systems and discrete time control systems based on the type of the signal used.
  - In continuous time control systems, all the signals are continuous in time. But, in discrete time control systems, there exists one or more discrete time signals.
- SISO and MIMO control systems:
  - Control Systems can be classified as SISO control systems and MIMO control systems based on the number of inputs and outputs present.
  - SISO (Single Input and Single Output) control systems have one input and one output. Whereas, MIMO (Multiple Inputs and Multiple Outputs) control systems have more than one input and more than one output.
- Open loop and closed loop control systems:
  - Control systems can be classified as open loop control systems and closed loop control systems based on the feedback path.
  - In open loop control systems, output is not fed-back to the input. So, the control action is independent of the desired output.

The following figure shows the block diagram of the open loop control system:

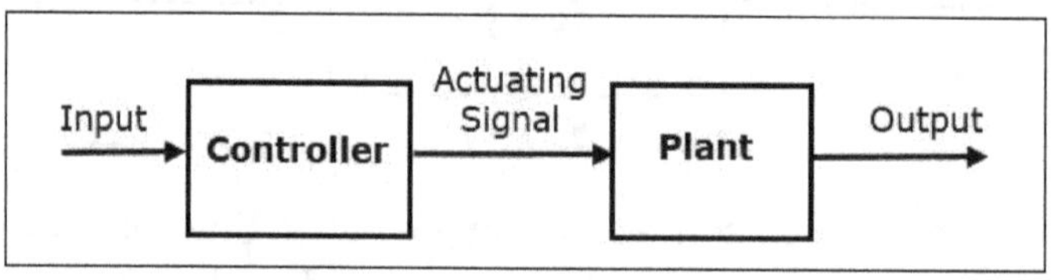

Here, an input is applied to a controller and it produces an actuating signal or controlling signal. This signal is given as an input to a plant or process which is to be controlled. So, the plant produces an output, which is controlled. The traffic lights control system which we discussed earlier is an example of an open loop control system.

In closed loop control systems, output is fed back to the input. So, the control action is dependent on the desired output.

The following figure shows the block diagram of negative feedback closed loop control system.

The error detector produces an error signal, which is the difference between the input and the feedback signal. This feedback signal is obtained from the block (feedback elements) by considering the output of the overall system as an input to this block. Instead of the direct input, the error signal is applied as an input to a controller.

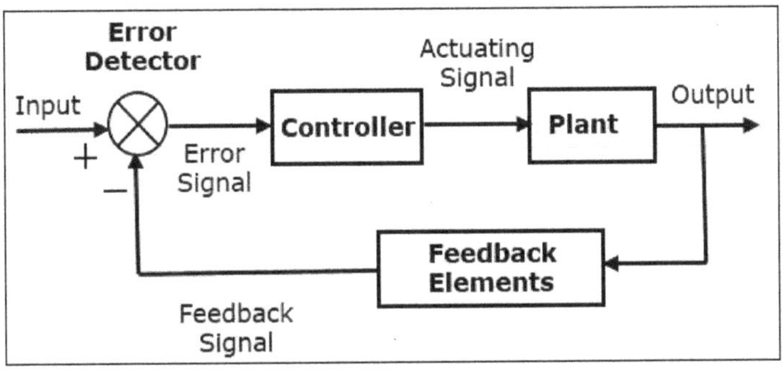

So, the controller produces an actuating signal which controls the plant. In this combination, the output of the control system is adjusted automatically till we get the desired response. Hence, the closed loop control systems are also called the automatic control systems. Traffic lights control system having sensor at the input is an example of a closed loop control system.

Open-loop System

The function of any electronic system is to automatically regulate the output and keep it within the systems desired input value or "set point". If the systems input changes for whatever reason, the output of the system must respond accordingly and change itself to reflect the new input value.

Likewise, if something happens to disturb the systems output without any change to the input value, the output must respond by returning back to its previous set value. In the past, electrical control systems were basically manual or what is called an Open-loop System with very few automatic control or feedback features built in to regulate the process variable so as to maintain the desired output level or value.

For example, an electric clothes dryer. Depending upon the amount of clothes or how wet they are, a user or operator would set a timer (controller) to say 30 minutes and at the end of the 30 minutes the drier will automatically stop and turn-off even if the clothes where still wet or damp.

In this case, the control action is the manual operator assessing the wetness of the clothes and setting the process (the drier) accordingly. So in this example, the clothes dryer would be an open-loop system as it does not monitor or measure the condition of the output signal, which is the dryness of the clothes. Then the accuracy of the drying process, or success of drying the clothes will depend on the experience of the user (operator).

However, the user may adjust or fine tune the drying process of the system at any time by increasing or decreasing the timing controllers drying time, if they think that the original drying process will not be met. For example, increasing the timing controller to 40 minutes to extend the drying process. Consider the following open-loop block diagram.

Open-loop Drying System

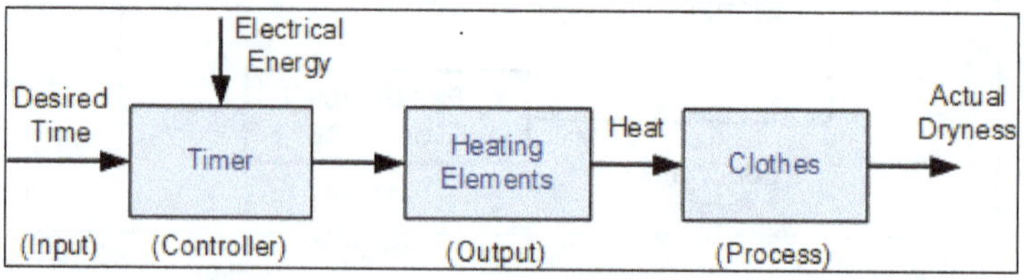

An Open-loop system, also referred to as non-feedback system, is a type of continuous control system in which the output has no influence or effect on the control action of the input signal. In other words, in an open-loop control system the output is neither measured nor "fed back" for comparison with the input. Therefore, an open-loop system is expected to faithfully follow its input command or set point regardless of the final result.

Also, an open-loop system has no knowledge of the output condition so cannot self-correct any errors it could make when the preset value drifts, even if this results in large deviations from the preset value.

Another disadvantage of open-loop systems is that they are poorly equipped to handle disturbances or changes in the conditions which may reduce its ability to complete the desired task. For example, the dryer door opens and heat is lost. The timing controller continues regardless for the full 30 minutes but the clothes are not heated or dried at the end of the drying process. This is because there is no information fed back to maintain a constant temperature.

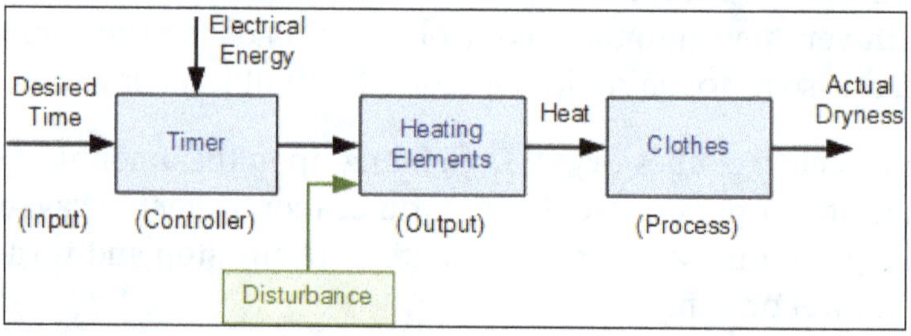

Then we can see that open-loop system errors can disturb the drying process and therefore requires extra supervisory attention of a user (operator). The problem with this anticipatory control approach is that the user would need to look at the process temperature frequently and take any corrective control action whenever the drying process deviated from its desired value of drying the clothes. This type of manual open-loop control which reacts before an error actually occurs is called Feed forward Control.

The objective of feed forward control, also known as predictive control, is to measure or predict any potential open-loop disturbances and compensate for them manually before the controlled variable deviates too far from the original set point. So for our simple example above, if the dryers door was open it would be detected and closed allowing the drying process to continue.

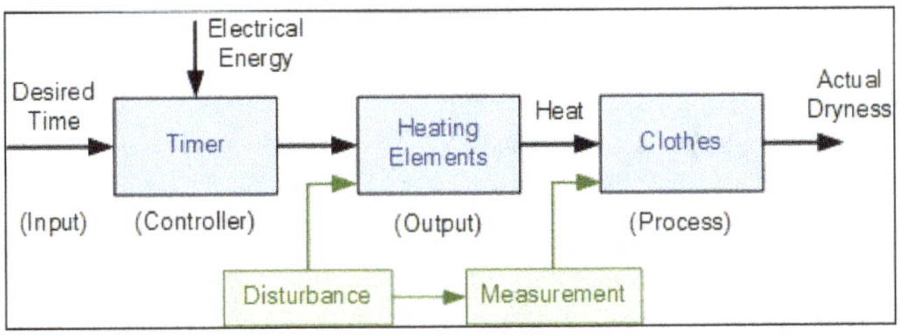

If applied correctly, the deviation from wet clothes to dry clothes at the end of the 30 minutes would be minimal if the user responded to the error situation (door open) very quickly. However, this feed forward approach may not be completely accurate if the system changes, for example the drop in drying temperature was not noticed during the 30 minute process.

Then we can define the main characteristics of an "Open-loop system" as being:

- There is no comparison between actual and desired values.
- An open-loop system has no self-regulation or control action over the output value.
- Each input setting determines a fixed operating position for the controller.
- Changes or disturbances in external conditions does not result in a direct output change (unless the controller setting is altered manually).

Any open-loop system can be represented as multiple cascaded blocks in series or a single block diagram with an input and output. The block diagram of an open-loop system shows that the signal path from input to output represents a linear path with no feedback loop and for any type of control system the input is given the designation θ_i and the output θ_o.

Generally, we do not have to manipulate the open-loop block diagram to calculate its actual transfer function. We can just write down the proper relationships or equations from each block diagram, and then calculate the final transfer function from these equations as shown.

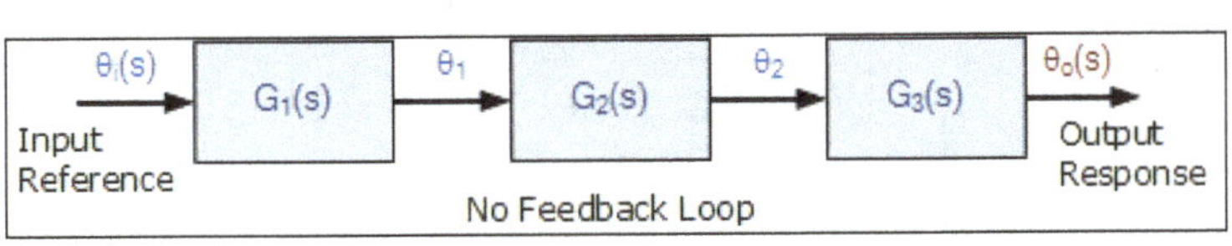

Open-loop System.

The Transfer Function of each block is therefore:

$$G_1 = \frac{\theta_1}{\theta_i},\ G_2 = \frac{\theta_2}{\theta_1},\ G_3 = \frac{\theta_o}{\theta_2},$$

The overall transfer function is given as:

$$G_1 \times G_2 \times G_3 = \frac{\not{\theta}_1}{\theta_i} \times \frac{\not{\theta}_2}{\not{\theta}_1} \times \frac{\theta_o}{\not{\theta}_2} \times \frac{\theta_o}{\theta_i},$$

Then the Open-loop Gain is given simply as:

$$\text{Gain,}(G) = \frac{\theta_o(s)}{\theta_i(s)}$$

When G represents the Transfer Function of the system or subsystem, it can be rewritten as: $G(s) = \theta_o(s)/\theta_i(s)$.

Open-loop control systems are often used with processes that require the sequencing of events with the aid of "ON-OFF" signals. For example a washing machines which requires the water to be switched "ON" and then when full is switched "OFF" followed by the heater element being switched "ON" to heat the water and then at a suitable temperature is switched "OFF", and so on.

This type of "ON-OFF" open-loop control is suitable for systems where the changes in load occur slowly and the process is very slow acting, necessitating infrequent changes to the control action by an operator.

Open-loop Motor Control

So for example, assume the DC motor controller as shown. The speed of rotation of the motor will depend upon the voltage supplied to the amplifier (the controller) by the potentiometer. The value of the input voltage could be proportional to the position of the potentiometer.

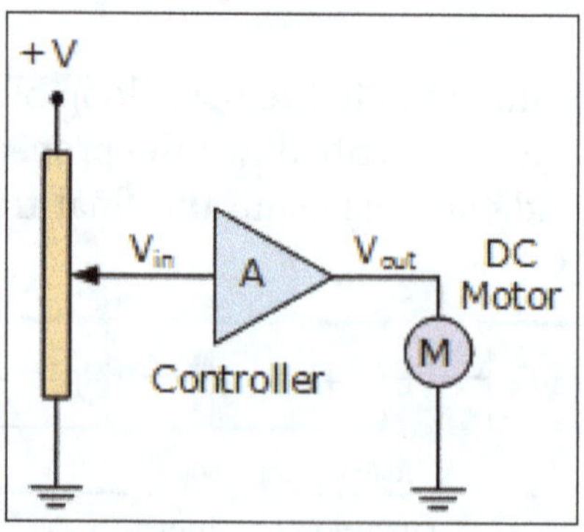

If the potentiometer is moved to the top of the resistance the maximum positive voltage will be supplied to the amplifier representing full speed. Likewise, if the potentiometer wiper is moved to the bottom of the resistance, zero voltage will be supplied representing a very slow speed or stop. Then the position of the potentiometers slider represents the input, θ_i which is amplified by the amplifier (controller) to drive the DC motor (process) at a set speed N representing the output, θ_o of the system. The motor will continue to rotate at a fixed speed determined by the position of the potentiometer.

As the signal path from the input to the output is a direct path not forming part of any loop, the overall gain of the system will the cascaded values of the individual gains from the potentiometer, amplifier, motor and load. It is clearly desirable that the output speed of the motor should be identical to the position of the potentiometer, giving the overall gain of the system as unity.

However, the individual gains of the potentiometer, amplifier and motor may vary over time with changes in supply voltage or temperature, or the motors load may increase representing external disturbances to the open-loop motor control system.

But the user will eventually become aware of the change in the systems performance (change in motor speed) and may correct it by increasing or decreasing the potentiometers input signal accordingly to maintain the original or desired speed.

The advantages of this type of "open-loop motor control" is that it is potentially cheap and simple to implement making it ideal for use in well-defined systems were the relationship between input and output is direct and not influenced by any outside disturbances. Unfortunately this type of open-loop system is inadequate as variations or disturbances in the system affect the speed of the motor. Then another form of control is required.

Closed-loop Systems

A Closed-loop Control System, also known as a feedback control system is a control system which uses the concept of an open loop system as its forward path but has one or more feedback loops (hence its name) or paths between its output and its input. The reference to "feedback", simply means that some portion of the output is returned "back" to the input to form part of the systems excitation.

Closed-loop systems are designed to automatically achieve and maintain the desired output condition by comparing it with the actual condition. It does this by generating an error signal which is the difference between the output and the reference input. In other words, a "closed-loop system" is a fully automatic control system in which its control action being dependent on the output in some way.

Suppose we used a sensor or transducer (input device) to continually monitor the temperature or dryness of the clothes and feed a signal relating to the dryness back to the controller as shown below.

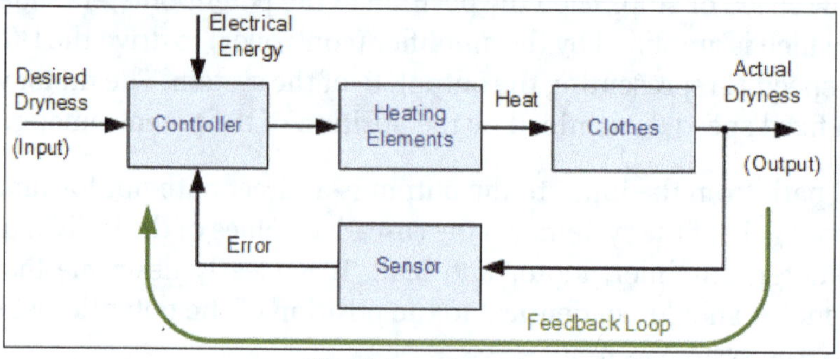

Closed-loop Control.

This sensor would monitor the actual dryness of the clothes and compare it with (or subtract it from) the input reference. The error signal (error = required dryness – actual dryness) is amplified by the controller, and the controller output makes the necessary correction to the heating system to reduce any error. For example if the clothes are too wet the controller may increase the temperature or drying time. Likewise, if the clothes are nearly dry it may reduce the temperature or stop the process so as not to overheat or burn the clothes, etc.

Then the closed-loop configuration is characterized by the feedback signal, derived from the sensor in our clothes drying system. The magnitude and polarity of the resulting error signal, would be directly related to the difference between the required dryness and actual dryness of the clothes.

Also, because a closed-loop system has some knowledge of the output condition, (via the sensor) it is better equipped to handle any system disturbances or changes in the conditions which may reduce its ability to complete the desired task.

For example, as before, the dryer door opens and heat is lost. This time the deviation in temperature is detected by the feedback sensor and the controller self-corrects the error to maintain a constant temperature within the limits of the preset value. Or possibly stops the process and activates an alarm to inform the operator.

In a closed-loop control system the error signal, which is the difference between the input signal and the feedback signal (which may be the output signal itself or a function of the output signal), is fed to the controller so as to reduce the systems error and bring the output of the system back to a desired value. In our case the dryness of the clothes. Clearly, when the error is zero the clothes are dry.

The term Closed-loop control always implies the use of a feedback control action in order to reduce any errors within the system, and its "feedback" which distinguishes the main differences between an open-loop and a closed-loop system. The accuracy of the output thus depends on the feedback path, which in general can be made very accurate and within electronic control systems and circuits, feedback control is more commonly used than open-loop or feed forward control.

Closed-loop systems have many advantages over open-loop systems. The primary advantage of a closed-loop feedback control system is its ability to reduce a system's sensitivity to external disturbances, for example opening of the dryer door, giving the system a more robust control as any changes in the feedback signal will result in compensation by the controller.

Then we can define the main characteristics of Closed-loop Control as being:

- To reduce errors by automatically adjusting the systems input.
- To improve stability of an unstable system.
- To increase or reduce the systems sensitivity.
- To enhance robustness against external disturbances to the process.
- To produce a reliable and repeatable performance.

Whilst a good closed-loop system can have many advantages over an open-loop control system, its main disadvantage is that in order to provide the required amount of control, a closed-loop system must be more complex by having one or more feedback paths. Also, if the gain of the controller is too sensitive to changes in its input commands or signals it can become unstable and start to oscillate as the controller tries to over-correct itself, and eventually something would break. So we need to "tell" the system how we want it to behave within some pre-defined limits.

Closed-loop Summing Points

For a closed-loop feedback system to regulate any control signal, it must first determine the error between the actual output and the desired output. This is achieved using a summing point, also referred to as a comparison element, between the feedback loop and the systems input. These summing points compare a systems set point to the actual value and produce a positive or negative error signal which the controller responds too, where: Error = Set point – Actual.

The symbol used to represent a summing point in closed-loop systems block-diagram is that of a circle with two crossed lines as shown. The summing point can either add signals together in which a Plus (+) symbol is used showing the device to be a "summer" (used for positive feedback), or it can subtract signals from each other in which case a Minus (–) symbol is used showing that the device is a "comparator" (used for negative feedback).

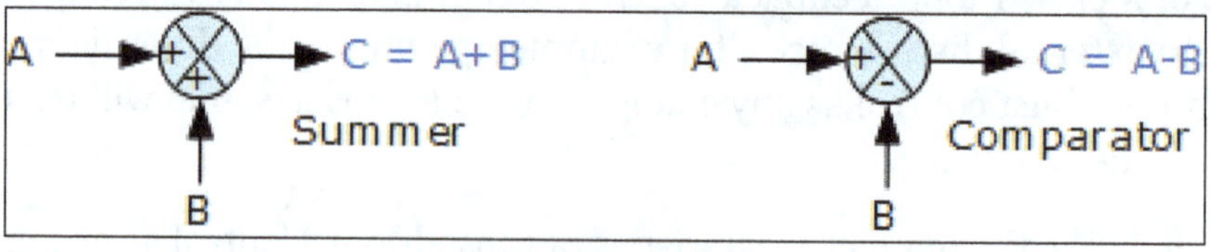

Summing Point Types.

Summing points can have more than one signal as inputs either adding or subtracting but only one output which is the algebraic sum of the inputs. Also the arrows indicate the direction of the signals. Summing points can be cascaded together to allow for more input variables to be summed at a given point.

Closed-loop System Transfer Function

The Transfer Function of any electrical or electronic control system is the mathematical relationship between the systems input and its output, and hence describes the behaviour of the system. Note also that the ratio of the output of a particular device to its input represents its gain. Then we can correctly say that the output is always the transfer function of the system times the input.

Typical Closed-loop System Representation

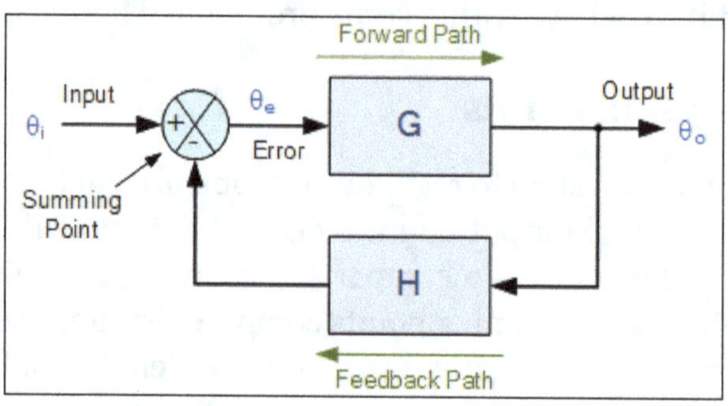

Where: block G represents the open-loop gains of the controller or system and is the forward path, and block H represents the gain of the sensor, transducer or measurement system in the feedback path.

To find the transfer function of the closed-loop system above, we must first calculate the output signal θ_o in terms of the input signal θ_i. To do so, we can easily write the equations of the given block-diagram as follows:

- The output from the system is equal to: Output = G x Error.

- The error signal, θ_e is also the input to the feed-forward block: G.
- The output from the summing point is equal to: Error = Input - H x Output.
- If H = 1 (unity feedback) then: The output from the summing point will be: Error (θ_e) = Input – Output.
- Eliminating the error term, then: The output is equal to: Output = G x (Input - H x Output).
- Therefore: G x Input = Output + G x H x Output

Rearranging the above gives us the closed-loop transfer function of:

$$\frac{\text{Output}}{\text{Iutput}} = \frac{\theta_o}{\theta_i} = \frac{G}{1+GH}$$

The above equation for the transfer function of a closed-loop system shows a Plus (+) sign in the denominator representing negative feedback. With a positive feedback system, the denominator will have a Minus (–) sign and the equation becomes: 1 - GH.

We can see that when H = 1 (unity feedback) and G is very large, the transfer function approaches unity as:

$$\frac{\text{Output}}{\text{Iutput}} \to 1$$

Also, as the systems steady state gain G decreases, the expression of: G/(1 + G) decreases much more slowly. In other words, the system is fairly insensitive to variations in the systems gain represented by G, and which is one of the main advantages of a closed-loop system.

Multi-loop Closed-loop System

Whilst our example above is of a single input, single output closed-loop system, the basic transfer function still applies to more complex multi-loop systems. Most practical feedback circuits have some form of multiple loop control, and for a multi-loop configuration the transfer function between a controlled and a manipulated variable depends on whether the other feedback control loops are open or closed.

Consider the multi-loop system below:

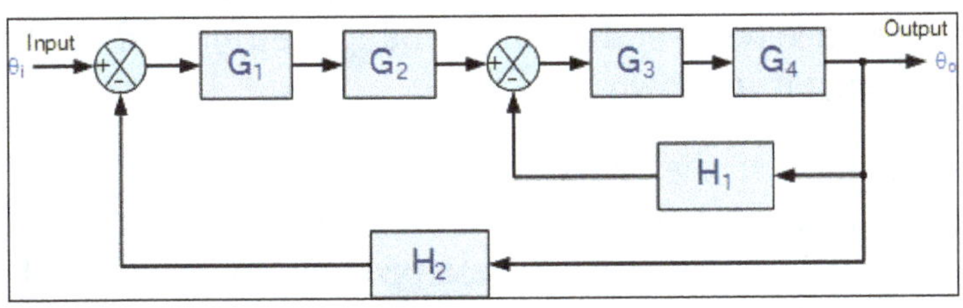

Any cascaded blocks such as G_1 and G_2 can be reduced, as well as the transfer function of the inner loop as shown:

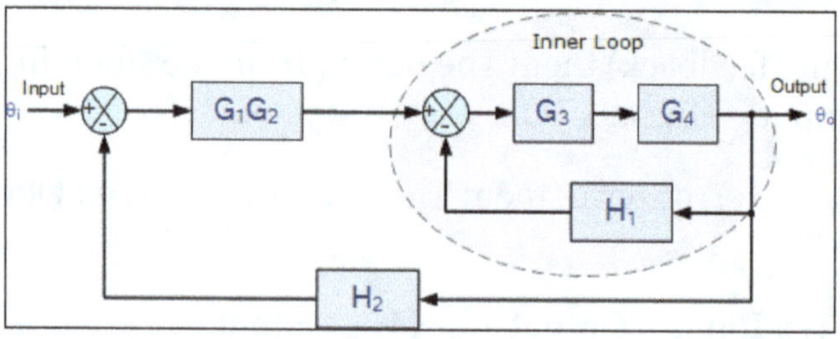

After further reduction of the blocks we end up with a final block diagram which resembles that of the previous single-loop closed-loop system.

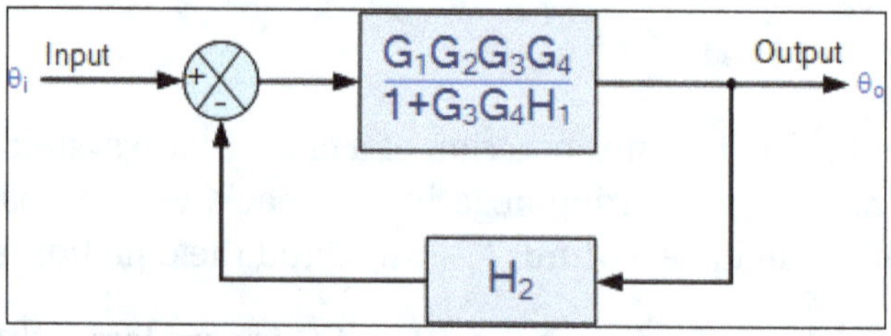

And the transfer function of this multi-loop system becomes:

$$\frac{\text{Output}}{\text{Iutput}} = \frac{\theta_o}{\theta_i} = \frac{G_1 G_2 G_3 G_4}{1 + G_3 G_4 H_1 + G_1 G_2 G_3 G_4 H_2}$$

Then we can see that even complex multi-block or multi-loop block diagrams can be reduced to give one single block diagram with one common system transfer function.

Closed-loop Motor Control

So how can we use Closed-loop Systems in Electronics. If we connected a speed measuring transducer, such as a tachometer to the shaft of the DC motor, we could detect its speed and send a signal proportional to the motor speed back to the amplifier. A tachometer, also known as a tacho-generator is simply a permanent-magnet DC generator which gives a DC output voltage proportional to the speed of the motor.

Then the position of the potentiometers slider represents the input, θ_i which is amplified by the amplifier (controller) to drive the DC motor at a set speed N representing the output, θ_o of the system, and the tachometer T would be the closed-loop back to the controller. The difference between the input voltage setting and the feedback voltage level gives the error signal as shown.

Any external disturbances to the closed-loop motor control system such as the motors load increasing would create a difference in the actual motor speed and the potentiometer input set point. This difference would produce an error signal which the controller would automatically respond too adjusting the motors speed. Then the controller works to minimize the error signal, with zero error indicating actual speed which equals set point.

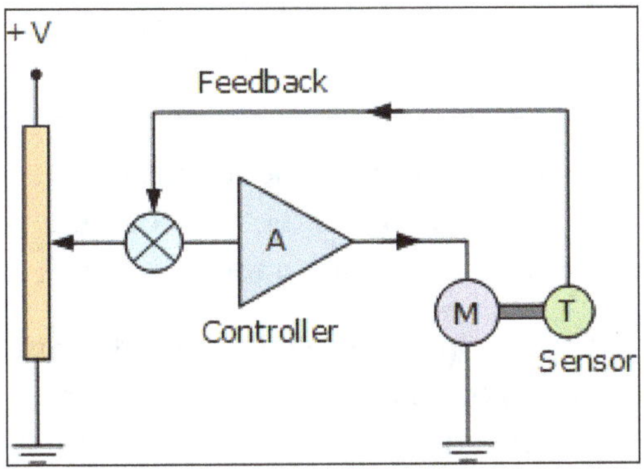

Closed-loop Motor Control.

Electronically, we could implement such a simple closed-loop tachometer-feedback motor control circuit using an operational amplifier (op-amp) for the controller as shown.

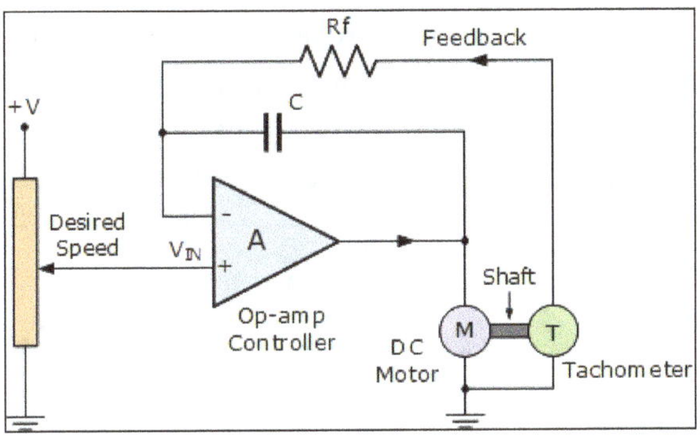

This simple closed-loop motor controller can be represented as a block diagram as shown:

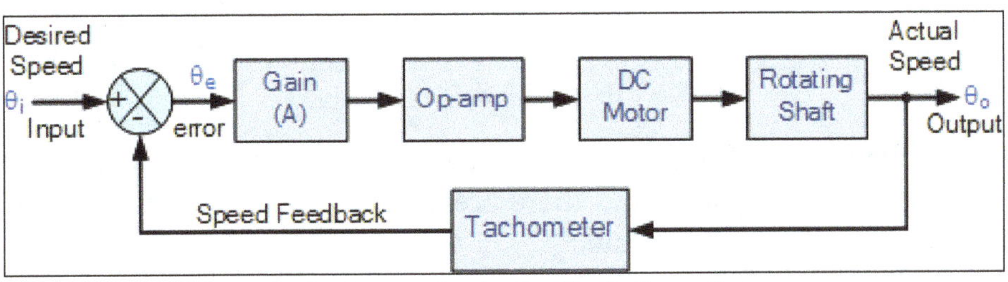

Block Diagram for the Feedback Controller.

A closed-loop motor controller is a common means of maintaining a desired motor speed under varying load conditions by changing the average voltage applied to the input from the controller. The tachometer could be replaced by an optical encoder or Hall-effect type positional or rotary sensor.

Feedback Systems

Feedback Systems process signals and as such are signal processors. The processing part of a feedback system may be electrical or electronic, ranging from a very simple to a highly complex circuits.

Simple analogue feedback control circuits can be constructed using individual or discrete components, such as transistors, resistors and capacitors, etc. or by using microprocessor-based and integrated circuits (IC's) to form more complex digital feedback systems.

Open-loop systems are just that, open ended, and no attempt is made to compensate for changes in circuit conditions or changes in load conditions due to variations in circuit parameters, such as gain and stability, temperature, supply voltage variations and/or external disturbances. But the effects of these "open-loop" variations can be eliminated or at least considerably reduced by the introduction of Feedback.

A feedback system is one in which the output signal is sampled and then fed back to the input to form an error signal that drives the system.

Feedback Systems are very useful and widely used in amplifier circuits, oscillators, process control systems as well as other types of electronic systems. But for feedback to be an effective tool it must be controlled as an uncontrolled system will either oscillate or fail to function. The basic model of a feedback system is given as:

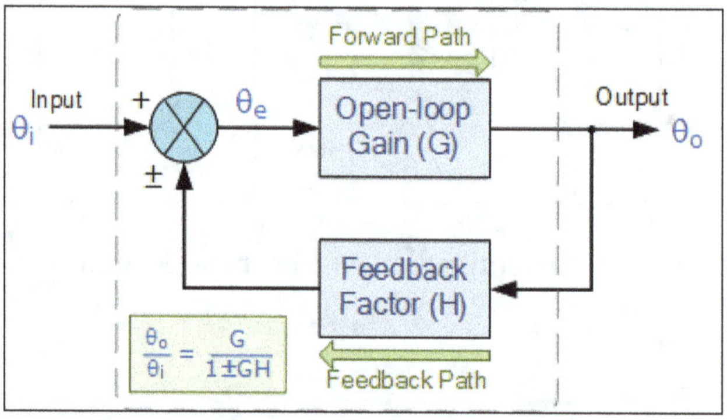

Feedback System Block Diagram Model.

This basic feedback loop of sensing, controlling and actuation is the main concept behind a feedback control system and there are several good reasons why feedback is applied and used in electronic circuits:

- Circuit characteristics such as the systems gain and response can be precisely controlled.
- Circuit characteristics can be made independent of operating conditions such as supply voltages or temperature variations.

- Signal distortion due to the non-linear nature of the components used can be greatly reduced.
- The Frequency Response, Gain and Bandwidth of a circuit or system can be easily controlled to within tight limits.

Classification of Feedback Systems

Thus far we have seen the way in which the output signal is "fed back" to the input terminal, and for feedback systems this can be of either, Positive Feedback or Negative Feedback. But the manner in which the output signal is measured and introduced into the input circuit can be very different leading to four basic classifications of feedback.

Based on the input quantity being amplified, and on the desired output condition, the input and output variables can be modelled as either a voltage or a current. As a result, there are four basic classifications of single-loop feedback system in which the output signal is fed back to the input and these are:

- Series-Shunt Configuration: Voltage in and Voltage out or Voltage Controlled Voltage Source (VCVS).
- Shunt-Shunt Configuration: Current in and Voltage out or Current Controlled Voltage Source (CCVS).
- Series-Series Configuration: Voltage in and Current out or Voltage Controlled Current Source (VCCS).
- Shunt-Series Configuration: Current in and Current out or Current Controlled Current Source (CCCS).

These names come from the way that the feedback network connects between the input and output stages as shown.

Series-Shunt Feedback Systems

Series-Shunt Feedback, also known as series voltage feedback, operates as a voltage-voltage controlled feedback system. The error voltage fed back from the feedback network is in series with the input. The voltage which is fed back from the output being proportional to the output voltage, V_o as it is parallel, or shunt connected.

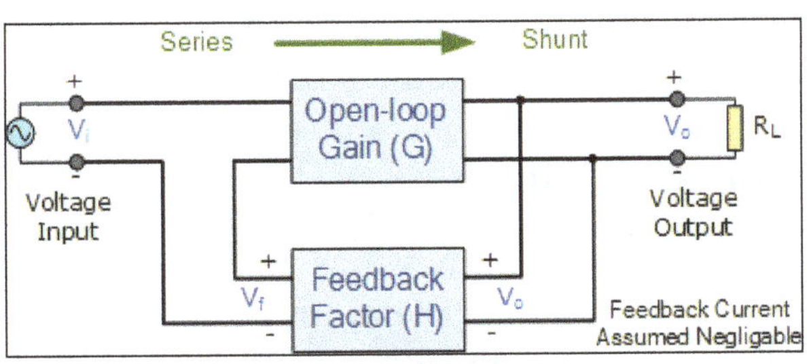

Series-Shunt Feedback System.

For the series-shunt connection, the configuration is defined as the output voltage, V_{out} to the input voltage, V_{in}. Most inverting and non-inverting operational amplifier circuits operate with series-shunt feedback producing what is known as a "voltage amplifier". As a voltage amplifier the ideal input resistance, R_{in} is very large, and the ideal output resistance, Rout is very small.

Then the "series-shunt feedback configuration" works as a true voltage amplifier as the input signal is a voltage and the output signal is a voltage, so the transfer gain is given as: $A_v = V_{out} \div V_{in}$. This quantity is dimensionless as its units are volts/volts.

Shunt-Series Feedback Systems

Shunt-Series Feedback, also known as shunt current feedback, operates as a current-current controlled feedback system. The feedback signal is proportional to the output current, I_o flowing in the load. The feedback signal is fed back in parallel or shunt with the input as shown.

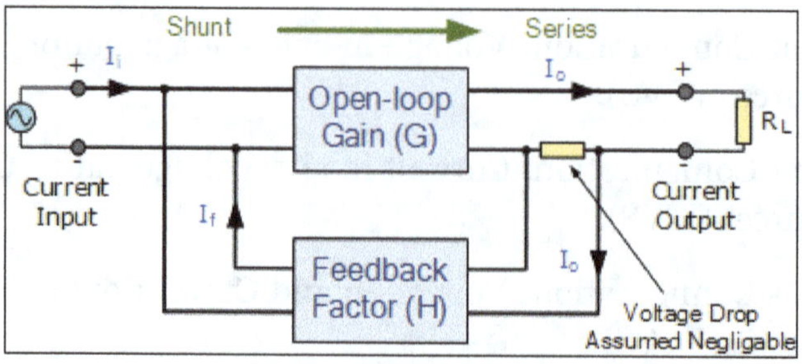

Shunt-Series Feedback System.

For the shunt-series connection, the configuration is defined as the output current, I_{out} to the input current, I_{in}. In the shunt-series feedback configuration the signal fed back is in parallel with the input signal and as such as the currents, not the voltages that add.

This parallel shunt feedback connection will not normally affect the voltage gain of the system, since for a voltage output a voltage input is required. Also, the series connection at the output increases output resistance, Rout while the shunt connection at the input decreases the input resistance, R_{in}.

Then the "shunt-series feedback configuration" works as a true current amplifier as the input signal is a current and the output signal is a current, so the transfer gain is given as: $A_i = I_{out} \div I_{in}$. This quantity is dimensionless as its units are amperes/amperes.

Series-Series Feedback Systems

Series-Series Feedback Systems, also known as series current feedback, operates as a voltage-current controlled feedback system. In the series current configuration the feedback error signal is in series with the input and is proportional to the load current,

I_{out}. Actually, this type of feedback converts the current signal into a voltage which is actually fed back and it is this voltage which is subtracted from the input.

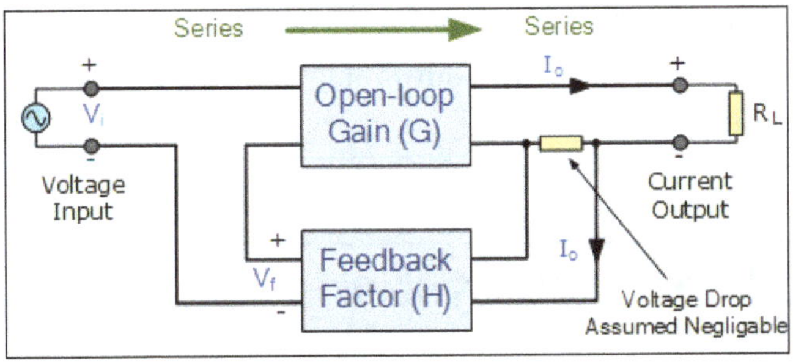

Series-Series Feedback System.

For the series-series connection, the configuration is defined as the output current, I_{out} to the input voltage, V_{in}. Because the output current, I_{out} of the series connection is fed back as a voltage, this increases both the input and output impedances of the system. Therefore, the circuit works best as a transconductance amplifier with the ideal input resistance, R_{in} being very large, and the ideal output resistance, Rout is also very large.

Then the "series-series feedback configuration" functions as transconductance type amplifier system as the input signal is a voltage and the output signal is a current. Then for a series-series feedback circuit the transfer gain is given as: $G_m = I_{out} \div V_{in}$.

Shunt-Shunt Feedback Systems

Shunt-Shunt Feedback Systems, also known as shunt voltage feedback, operates as a current-voltage controlled feedback system. In the shunt-shunt feedback configuration the signal fed back is in parallel with the input signal. The output voltage is sensed and the current is subtracted from the input current in shunt, and as such as the currents, not the voltages that subtract.

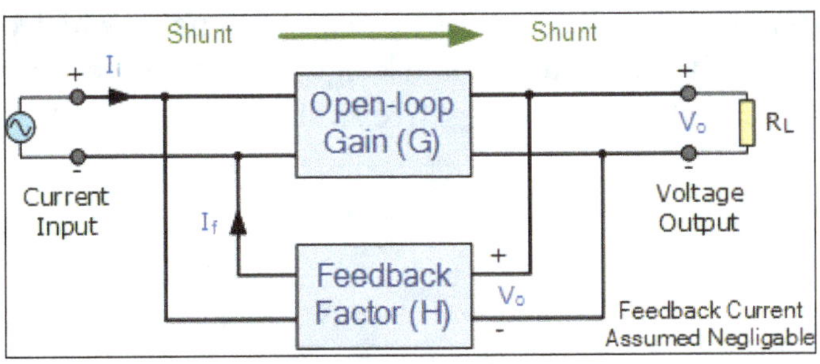

Shunt-Shunt Feedback System.

For the shunt-shunt connection, the configuration is defined as the output voltage, V_{out} to the input current, I_{in}. As the output voltage is fed back as a current to a current-driven input port, the shunt connections at both the input and output terminals reduce the input and output impedance. Therefore the system works best as a transresistance system with the ideal input resistance, R_{in} being very small, and the ideal output resistance, Rout also being very small.

Then the shunt voltage configuration works as transresistance type voltage amplifier as the input signal is a current and the output signal is a voltage, so the transfer gain is given as: $R_m = V_{out} \div I_{in}$.

Negative Feedback Systems

Feedback is the process by which a fraction of the output signal, either a voltage or a current, is used as an input. If this feedback fraction is opposite in value or phase ("anti-phase") to the input signal, then the feedback is said to be Negative Feedback, or degenerative feedback.

Negative feedback opposes or subtracts from the input signals giving it many advantages in the design and stabilisation of control systems. For example, if the systems output changes for any reason, then negative feedback affects the input in such a way as to counteract the change.

Feedback reduces the overall gain of a system with the degree of reduction being related to the systems open-loop gain. Negative feedback also has effects of reducing distortion, noise, sensitivity to external changes as well as improving system bandwidth and input and output impedances. Feedback in an electronic system, whether negative feedback or positive feedback is unilateral in direction. Meaning that its signals flow one way only from the output to the input of the system. This then makes the loop gain, G of the system independent of the load and source impedances.

As feedback implies a closed-loop system it must therefore have a summing point. In a negative feedback system this summing point or junction at its input subtracts the feedback signal from the input signal to form an error signal, β which drives the system. If the system has a positive gain, the feedback signal must be subtracted from the input signal in order for the feedback to be negative as shown.

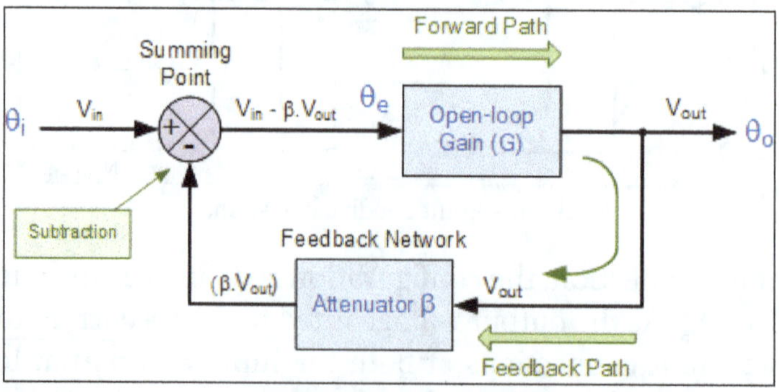

Negative Feedback Circuit.

The circuit represents a system with positive gain, G and feedback, β. The summing junction at its input subtracts the feedback signal from the input signal to form the error signal V_{in} - βG, which drives the system.

Then using the basic closed-loop circuit above we can derive the general feedback equation as being:

Negative Feedback Equation:

$$\text{System Gain, } G = \frac{V_{out}}{V_{in}} \quad G = \text{open loop voltage gain}$$

$$G \times V_n = V_{out}$$

$$G(V_{in} - \beta V_{out}) = V_{out} \qquad \beta \text{ is the feedback fraction}$$

$$G.V_{in} - \beta.G.V_{out} = V_{out} \qquad \beta G = \text{the loop gain}$$

$$G.V_{in} = V_{out}(1+\beta G) \qquad 1+\beta G = \text{the feedback factor}$$

$$\therefore \frac{V_{out}}{V_{in}} = Gv = \frac{G}{1+\beta G} \qquad Gv = \text{closed loop voltage gain}$$

We see that the effect of the negative feedback is to reduce the gain by the factor of: 1 + βG. This factor is called the "feedback factor" or "amount of feedback" and is often specified in decibels (dB) by the relationship of 20 log (1+ βG).

Effects of Negative Feedback

If the open-loop gain, G is very large, then βG will be much greater than 1, so that the overall gain of the system is roughly equal to 1/β. If the open-loop gain decreases due to frequency or the effects of system ageing, providing that βG is still relatively large, the overall system gain does not change very much. So negative feedback tends to reduce the effects of gain change giving what is generally called "gain stability".

Examples of Negative Feedback

Example: A system has a gain of 80dB without feedback. If the negative feedback fraction is 1/50th. Calculate the closed-loop gain of the system in dB with the addition of negative feedback.

$$80\text{dB} = 20\log(G)$$

$$\therefore G = \text{anti}\log 10^{80/20} = 10{,}000$$

$$Gv = \frac{G}{1+G\beta} = \frac{10{,}000}{1+10{,}000.\left(\frac{1}{50}\right)} = 49.75$$

$$\therefore Gv_{(dB)} = 20\text{LOG}(49.75) = 34\text{dB}$$

Then we can see that the system has a loop gain of 10,000 and a closed-loop gain of 34dB.

Example: If after 5 years the loop gain of the system without negative feedback has fallen to 60dB and the feedback fraction has remained constant at 1/50th. Calculate the new closed-loop gain value of the system.

$$60\text{dB} = 20\log(G)$$

$$\therefore G = \text{anti}\log 10^{60/20} = 1{,}000$$

$$Gv = \frac{G}{1+G\beta} = \frac{1{,}000}{1+1{,}000.\left(\frac{1}{50}\right)} = 47.6$$

$$\therefore Gv_{(dB)} = 20\text{LOG}(47.6) = 33.5\text{dB}$$

Then we can see from the two examples that without feedback, after 5 years of use the systems gain has fallen from 80dB down to 60dB, (10,000 to 1,000) a drop in open loop gain of about 25%.

However with the addition of negative feedback the systems gain has only fallen from 34dB to 33.5dB, a reduction of less than 1.5%, which proves that negative feedback gives added stability to a systems gain. Therefore we can see that by applying negative feedback to a system greatly reduces its overall gain compared to its gain without feedback.

The systems gain without feedback can be very large but not precise as it may change from one system device to the next, then it is possible to design a system with sufficient open-loop gain that, after the negative feedback has been added, the overall gain matches the desired value.

Also, if the feedback network is constructed from passive elements having stable characteristics, the overall gain becomes very steady and unaffected by variation in the systems inherent open-loop gain.

Negative Feedback in Operational Amplifiers

Operational amplifiers (op-amps) are the most commonly used type of linear integrated circuit but they have a very high gain. The open-loop voltage gain, A_{VOL}, of a standard 741 op-amp is its voltage gain when there is no negative feedback applied and the open-loop voltage gain of an op-amp is the ratio of its output voltage, V_{out}, to its differential input voltage, V_{in}, (V_{out}/V_{in}).

The typical value of A_{VOL} for a 741 op-amp is more than 200,000 (106dB). So an input voltage signal of only 1mV, would result in an output voltage of over 200 volts forcing the output immediately into saturation. Obviously this high open-loop voltage gain needs to be controlled in some way, and we can do just that by using negative feedback.

The use of negative feedback can significantly improve the performance of an operational amplifier and any op-amp circuit that does not use negative feedback is considered too unstable to be useful. But how can we use negative feedback to control an op-amp. Well consider the circuit below of a Non-inverting Operational Amplifier.

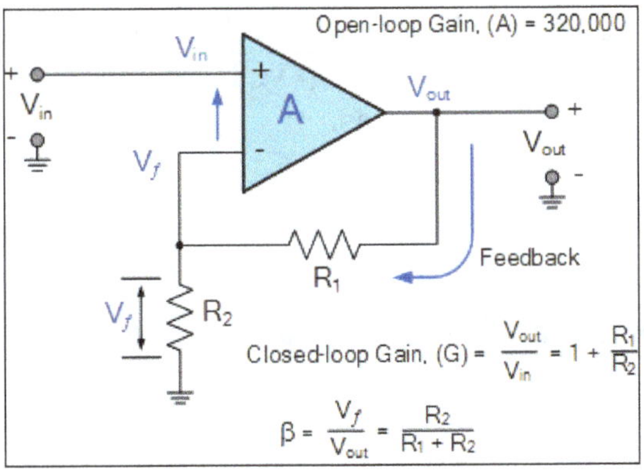

Non-inverting Op-amp Circuit.

Example: An operational amplifier with an open-loop voltage gain, A_{VOL} of 320,000 without feedback is to be used as a non-inverting amplifier. Calculate the values of the feedback resistances, R_1 and R_2 required to stabilise the circuit with a closed loop gain of 20.

The generalised closed-loop feedback equation we derived above is given as:

$$G = \frac{A}{1+\beta A}$$

By rearranging the feedback formula we get a feedback fraction, β of:

$$G = (1+\beta A) = A$$

$$1+\beta A = \frac{A}{G}$$

$$\beta A = \frac{A}{G} - 1$$

$$\therefore \beta = \frac{1}{G} - \frac{1}{A}$$

Then putting the values of: A = 320,000 and G = 20, into the above equation we get the value of β as:

$$\beta = \frac{1}{G} - \frac{1}{A}$$

$$\beta = \frac{1}{20} - \frac{1}{320{,}000} = 0.05$$

Because in this case the open-loop gain of the op-amp is very high (A = 320,000), the feedback fraction, β will be roughly equal to the reciprocal of the closed-loop gain 1/G only as the value of 1/A will be incredibly small. Then β (the feedback fraction) is equal to 1/20 = 0.05.

As the resistors, R_1 and R_2 form a simple series-voltage potential divider network across the non-inverting amplifier, the closed-loop voltage gain of the circuit will be determined by the ratios of these resistances as:

$$\beta = \frac{V_f}{V_{OUT}} = \frac{R_2}{R_1 + R_2}$$

If we assume resistor R_2 has a value of 1,000Ω, or 1kΩ, then the value of resistor R_1 will be:

$$\beta = \frac{R_2}{R_1 + R_2}$$

$$\therefore R_1 = \frac{R_2 - \beta R_2}{\beta}$$

$$= \frac{1000 - (0.05 \times 1000)}{0.05} = 19{,}000\Omega = 19k\Omega$$

Then for the non-inverting amplifier circuit about to have a closed-loop gain of 20, the values of the negative feedback resistors required will be in this case, R_1 = 19kΩ and R_2 = 1kΩ, giving us a non-inverting amplifier circuit of:

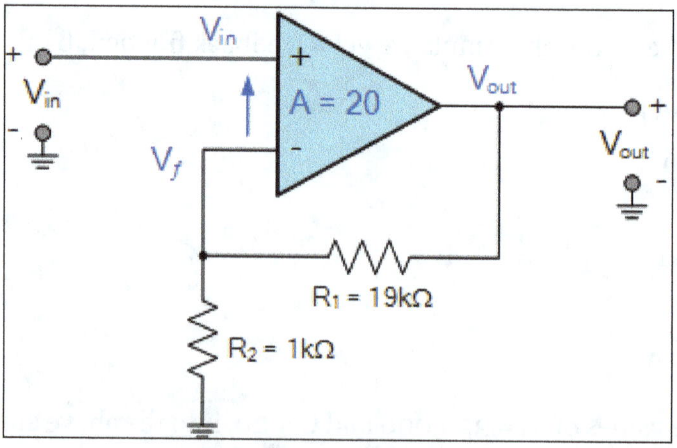

Non-inverting Op-amp Circuit.

There are many advantages to using feedback within a systems design, but the main advantages of using Negative Feedback in amplifier circuits is to greatly improve their stability, better tolerance to component variations, stabilisation against DC drift as well as increasing the amplifiers bandwidth.

Examples of negative feedback in common amplifier circuits include the resistor R_f in op-amp circuits, resistor, R_S in FET based amplifiers and resistor, R_E in bipolar transistor (BJT) amplifiers.

PID Control

A proportional–integral–derivative controller (PID controller or three-term controller) is a control loop mechanism employing feedback that is widely used in industrial control systems and a variety of other applications requiring continuously modulated control. A PID controller continuously calculates an error value e(t) as the difference between a desired setpoint (SP) and a measured process variable (PV) and applies a correction based on proportional, integral, and derivative terms (denoted *P*, *I*, and *D* respectively), hence the name.

In practical terms it automatically applies accurate and responsive correction to a control function. An everyday example is the cruise control on a car, where ascending a hill would lower speed if only constant engine power were applied. The controller's PID algorithm restores the measured speed to the desired speed with minimal delay and overshoot by increasing the power output of the engine.

The first theoretical analysis and practical application was in the field of automatic steering systems for ships, developed from the early 1920s onwards. It was then used for automatic process control in the manufacturing industry, where it was widely implemented in pneumatic, and then electronic, controllers. Today the PID concept is used universally in applications requiring accurate and optimised automatic control.

Fundamental Operation

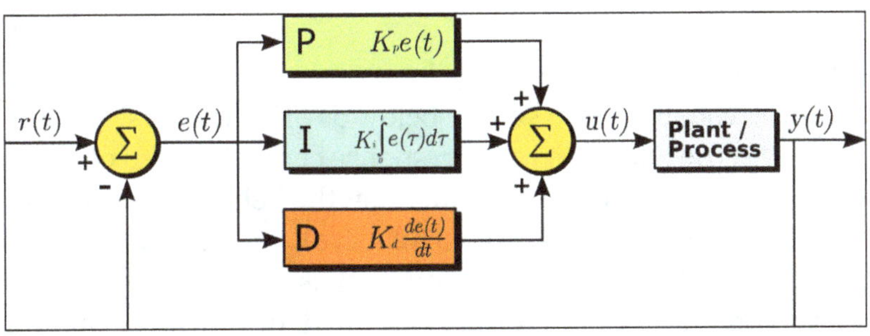

A block diagram of a PID controller in a feedback loop. $r(t)$ is the desired process value or setpoint (SP), and $y(t)$ is the measured process value (PV).

The distinguishing feature of the PID controller is the ability to use the three control terms of proportional, integral and derivative influence on the controller output to apply accurate and optimal control. The block diagram on the right shows the principles of how these terms are generated and applied. It shows a PID controller, which continuously calculates an error value $e(t)$ as the difference between a desired setpoint $SP = r(t)$ and a measured process variable $PV = y(t)$, and applies a correction based on proportional, integral, and derivative terms. The controller attempts to minimize the error over time by adjustment of a control variable $u(t)$, such as the opening of a control valve, to a new value determined by a weighted sum of the control terms.

In this model:

- Term P is proportional to the current value of the SP – PV error e(t). For example, if the error is large and positive, the control output will be proportionately large and positive, taking into account the gain factor "K". Using proportional control alone will result in an error between the setpoint and the actual process value, because it requires an error to generate the proportional response. If there is no error, there is no corrective response.
- Term I accounts for past values of the SP – PV error and integrates them over time to produce the I term. For example, if there is a residual SP – PV error after the application of proportional control, the integral term seeks to eliminate the residual error by adding a control effect due to the historic cumulative value of the error. When the error is eliminated, the integral term will cease to grow. This will result in the proportional effect diminishing as the error decreases, but this is compensated for by the growing integral effect.
- Term D is a best estimate of the future trend of the SP – PV error, based on its current rate of change. It is sometimes called "anticipatory control", as it is effectively seeking to reduce the effect of the SP – PV error by exerting a control influence generated by the rate of error change. The more rapid the change, the greater the controlling or dampening effect.

Tuning

The balance of these effects is achieved by loop tuning to produce the optimal control function. The tuning constants are shown below as "K" and must be derived for each control application, as they depend on the response characteristics of the complete loop external to the controller. These are dependent on the behaviour of the measuring sensor, the final control element (such as a control valve), any control signal delays and the process itself. Approximate values of constants can usually be initially entered knowing the type of application, but they are normally refined, or tuned, by "bumping" the process in practice by introducing a setpoint change and observing the system response.

Control Action

The mathematical model and practical loop above both use a "direct" control action for all the terms, which means an increasing positive error results in an increasing positive control output for the summed terms to apply correction. However, the output is called "reverse" acting if it is necessary to apply negative corrective action. For instance, if the valve in the flow loop was 100–0% valve opening for 0–100% control output – meaning that the controller action has to be reversed. Some process control schemes and final control elements require this reverse action. An example would be a valve for cooling water, where the fail-safe mode, in the case of loss of signal, would be 100% opening of the valve; therefore 0% controller output needs to cause 100% valve opening.

References

- Control-systems-introduction, control-systems: tutorialspoint.com, Retrieved 18 March, 2020
- Open-loop-system, systems: electronics-tutorials.ws, Retrieved 14 April, 2020
- Closed-loop-system, systems: electronics-tutorials.ws, Retrieved 20 July, 2020
- Controllers-and-their-characterstics diy-electronics-devices-54829: brighthubengineering.com, Retrieved 28 February, 2020
- Derivative-controller-control-system: myclassbook.org, Retrieved 13 August, 2020
- Integral-control, tutorials-language-pbasic-pid-control: learn.parallax.com, Retrieved 10 January, 2020
- Lipták, Béla G. (2003). Instrument Engineers' Handbook: Process control and optimization (4th ed.). CRC Press. p. 108. ISBN 0-8493-1081-4

Permissions

All chapters in this book are published with permission under the Creative Commons Attribution Share Alike License or equivalent. Every chapter published in this book has been scrutinized by our experts. Their significance has been extensively debated. The topics covered herein carry significant information for a comprehensive understanding. They may even be implemented as practical applications or may be referred to as a beginning point for further studies.

We would like to thank the editorial team for lending their expertise to make the book truly unique. They have played a crucial role in the development of this book. Without their invaluable contributions this book wouldn't have been possible. They have made vital efforts to compile up to date information on the varied aspects of this subject to make this book a valuable addition to the collection of many professionals and students.

This book was conceptualized with the vision of imparting up-to-date and integrated information in this field. To ensure the same, a matchless editorial board was set up. Every individual on the board went through rigorous rounds of assessment to prove their worth. After which they invested a large part of their time researching and compiling the most relevant data for our readers.

The editorial board has been involved in producing this book since its inception. They have spent rigorous hours researching and exploring the diverse topics which have resulted in the successful publishing of this book. They have passed on their knowledge of decades through this book. To expedite this challenging task, the publisher supported the team at every step. A small team of assistant editors was also appointed to further simplify the editing procedure and attain best results for the readers.

Apart from the editorial board, the designing team has also invested a significant amount of their time in understanding the subject and creating the most relevant covers. They scrutinized every image to scout for the most suitable representation of the subject and create an appropriate cover for the book.

The publishing team has been an ardent support to the editorial, designing and production team. Their endless efforts to recruit the best for this project, has resulted in the accomplishment of this book. They are a veteran in the field of academics and their pool of knowledge is as vast as their experience in printing. Their expertise and guidance has proved useful at every step. Their uncompromising quality standards have made this book an exceptional effort. Their encouragement from time to time has been an inspiration for everyone.

The publisher and the editorial board hope that this book will prove to be a valuable piece of knowledge for students, practitioners and scholars across the globe.

Index

www.ingramcontent.com/pod-product-compliance
Lightning Source LLC
LaVergne TN
LVHW080632210926
848485LV00003B/252